Supporting open minds since 2005

Applications of RNA-Seq in Biology and Medicine
http://dx.doi.org/10.5772/intechopen.91555
Edited by Irina Vlasova-St. Louis

Contributors
Venkateswara R. Sripathi, Varsha C. Anche, Zachary B. Gossett, Lloyd T. Walker, Ramya Ramadoss, Rajkumar Krishnan, Lekshmy Jayan, Priyadharini Shankaran, Richa Priydarshini, Karthik Krishnan, Rashmi Niranjan, Bhassu Subha, Sudhesh Dev Sareshma, Irina Vlasova-St. Louis, Andrew Gorzalski, Mark Pandori

First published in London, United Kingdom, 2021 by IntechOpen
IntechOpen is the global imprint of INTECHOPEN LIMITED, registered in England and Wales, registration number: 11086078, 5 Princes Gate Court, London, SW7 2QJ, United Kingdom
Printed in Croatia

British Library Cataloguing-in-Publication Data
A catalogue record for this book is available from the British Library

Additional hard and PDF copies can be obtained from orders@intechopen.com

Applications of RNA-Seq in Biology and Medicine
Edited by Irina Vlasova-St. Louis
p. cm.
Print ISBN 978-1-83962-686-9
Online ISBN 978-1-83962-815-3
eBook (PDF) ISBN 978-1-83962-816-0

For EU product safety concerns:
IN TECH d.o.o., Prolaz Marije Krucifikse Kozulić 3, 51000 Rijeka, Croatia, info@intechopen.com or visit our website at intechopen.com.

Meet the editor

Dr. Vlasova-St. Louis earned her MD and Ph.D. from Ural State Medical Academy, Russia. She completed her postdoctoral training at the University of Minnesota, USA, and a fellowship sponsored by the Lymphoma Research Foundation. She served as an assistant professor at the Department of Medicine, University of Minnesota. Dr. Vlasova-St. Louis has expertise in several biological disciplines including infectious diseases, immunology, and bioinformatics. By integrating state-of-the-art techniques such as next-generation sequencing, she made numerous biomedical discoveries studying normal and pathological conditions at the molecular, cellular, and organismal levels. Currently, Dr. St. Louis is a COVID-19 Associate, sponsored by the Association of Public Health Laboratories and the Center for Disease Control and Prevention. She leads the molecular surveillance program of novel SARS-CoV-2 variants.

Contents

Preface

The advent of next-generation sequencing along with the development of bioinformatics tools has opened avenues to explore this technology in numerous fields of biomedical research. This book evaluates and comprehensively summarizes the scientific findings that have been achieved through RNA-sequencing (RNA-Seq) technology.

RNA-Seq allows us to accurately capture all subtypes of RNA molecules, in any sequenced organism or single-cell type, under different experimental conditions. RNA-Seq transcriptome profiling of healthy and diseased tissues allows for understanding the alterations in cellular phenotypes through the expression of differentially spliced RNA isoforms. Assessment of gene expression by RNA-Seq provides new insight into host response to pathogens, drugs, allergens, and other environmental triggers.

RNA-Seq becomes even more powerful when combined with other assays. Merging genomics and transcriptomic profiling provides novel information underlying causative DNA mutations and the cellular effects of genetic variants caused by SNPs, indels, and more. Combining RNA-Seq with immunoprecipitation and cross-linking techniques is a clever multi-omics strategy assessing transcriptional, posttranscriptional, and posttranslational levels of gene expression regulation.

The optimization of RNA-Seq technology will allow countless opportunities in our pursuit of achieving the goals of individualized medicine and public health emergency responses.

Irina Vlasova-St. Louis, MD, Ph.D.
COVID-19 Laboratory Associate Program,
Nevada State Public Health Laboratory,
Association of Public Health Laboratories,
Center for Disease Control and Prevention,
Reno, NV, USA

Chapter 1

Introductory Chapter: Applications of RNA-Seq Diagnostics in Biology and Medicine

Irina Vlasova-St. Louis

1. Introduction

The advent of next generation sequencing along with the development of bioinformatics tools has opened avenues to explore this technology in numerous fields of biomedical research (**Figure 1**). This book evaluates and comprehensively summarizes the scientific findings which have been achieved through RNA-Sequencing technology (RNA-Seq). RNA-Seq allows accurate capture of all RNA molecule subtypes; in any sequenced organism or single-cell type, under various experimental conditions. Coding and noncoding RNA types of the gene can be analyzed as part of discovery research and diagnosis of diseases. RNA-Seq transcriptome profiling of both healthy and diseased tissues provides an understanding of the alterations, in cellular phenotypes, through the expression of RNA isoforms. Assessment of gene expression, by RNA-Seq, provides new insight into host response to pathogens, drugs, allergens, and other environmental triggers [1].

RNA-sequencing becomes even more powerful when combined with other assays. Merging genomic and transcriptomic profiling provides novel information about underlying causative DNA mutations and the cellular effects of genetic variants caused by single nucleotide polymorphism (SNPs), indels, etc. Combining RNA-Seq with, proteome evaluation, immunoprecipitation, and cross-linking techniques is a clever multi-omics strategy to assess transcriptional, posttranscriptional, and posttranslational levels of gene expression regulation. The optimization of

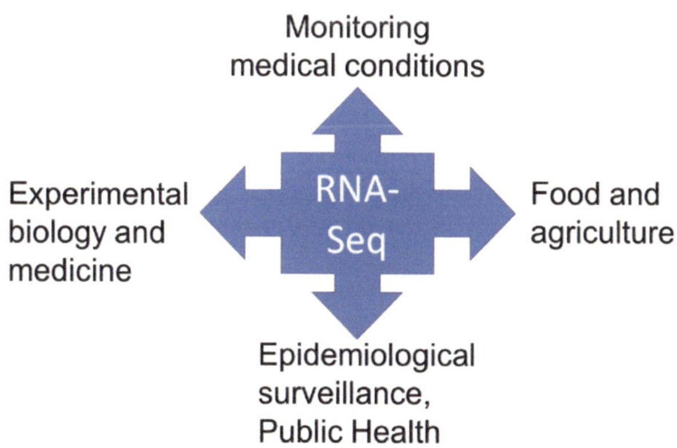

Figure 1.
Applications of RNA-Seq in biology and medicine. Applications of RNA-Seq discussed in this book.

RNA-Seq technology can provide countless opportunities in our pursuit of achieving the goals of systems biology and medicine, as described further in this chapter.

2. RNA sequencing in malignant and non-malignant disorders

RNA sequencing is a commercially available precision cancer diagnostic test [2]. No two tumors are alike. It is therefore imperative for cancer patients to have comprehensive testing of their tumors at the transcriptional and posttranscriptional levels [3]. Understanding the molecular features of cancer when the diagnosis is made allows oncologists to determine an optimal treatment path [4]. When making decisions about an individualized treatment many patients now understand the significance of having the most advanced molecular diagnostics available. Many companies (e.g., Caris Life Sciences) offer unique precision diagnostics services that are designed to maximize access to clinical trials, thus, offering patients a way toward novel treatments that are beyond the standard of care [5].

An increasing role of RNA sequencing in the detection of potentially malignant oral disorders, such as leukoplakia, lichen planus, or oral submucous fibrosis has been recently acknowledged. Transcriptome analysis of dysplastic tissues allows estimating of the rate of progression, of the pre-malignant conditions. Such estimation allows for individualized patient prognosis. The utilities for sequencing of oral, gut, and skin microbiota are becoming increasingly obvious. The steps in dual RNA sequencing are being continuously discussed, and RNA sequencing methodologies are being continuously improved [6].

3. Insights into normal microbiota and biofilms from metagenomics and transcriptomic approaches

The history of mapping the transcriptome via high throughput RNA sequencing methodologies is fascinating, with the earliest papers describing the term 'RNA-Seq' as pertaining to the yeast transcriptome [7].

RNA sequencing has become indispensable in metagenomics and meta- transcriptomics research, on human microbiota and biofilms. Our understanding of the magnitude and diversity of the species present within the gut, skin, and mucous microbiomes has increased dramatically in recent years. Meta- transcriptomic of 16s rRNA of human microbiome revealed exciting implications of altered biofilm maturation and dysbiosis, for various human diseases [8]. In this respect, a systems biology approach provides a larger picture of the molecules involved, in each system state or condition. Connecting RNA-Seq technologies with other multi- omics technologies allows for the identification, and selection, of key molecules and biomarkers of physiological as well as pathological processes [9]. The advancement of multi- omics technologies broaden the whole-system understanding of commensal and pathogenic microbiota [10]. It is very likely that we have still only scratched the surface of the plethora bacteria, fungi and viruses present within the human microbiome [11].

4. RNA-Seq profiling of host-pathogen interactions

Extraordinary effort by the scientific community has led to the development of various RNA-Seq procedures, including single-cell RNA-Seq and dual host-pathogen RNA-Seq for biology and medicine [12, 13]. Selection of specific RNA species for research can be carried out either by enriching transcripts expressing poly- adenylated tails (usually for mRNA profiling), or by removing the abundant ribosomal RNAs or globin RNAs [14]. However, these techniques still face many challenges when using RNA-Seq profiling in assessing host-pathogen interaction networks. Various RNA-Seq chemistries such as pyro- sequencing, sequencing by hybridization, and sequencing by synthesis of RNA molecules converted into cDNAs, brings forth a more functional and integrated view of expressed genes. Recently developed, and commercialized, nanopore-based sequencing allows direct RNA sequencing detection based on a unique fluctuation in ionic current while nucleotides pass through nano- channels [15]. These advances in medical sequencing methodologies enable deciphering the transcriptome architectures of human immunodeficiency virus (HIV), as well as severe acute respiratory syndrome coronavirus 2 (SARS-CoV-2). In turn, this led to the rapid establishment of viral sequencing-based genomic surveillance world-wide. Owing to RNA-Seq technologies we can now perform nearly real-time phylogenetic studies, and genomic epidemiology surveillance of novel SARS-CoV-2 variants.

Bioinformatics efforts significantly improve genomic annotations of novel RNA isoforms through examination of translated, untranslated, differentially spliced regions, or the allele-specific RNA expressions [16–18]. Bioinformatics characterization of pathological host immune gene expression allowed for discoveries of novel biomarkers of immune reconstitution inflammatory syndrome (IRIS), in the HIV-infected population [19–21]. Additionally, severe illnesses that begin with uncontrolled overexpression of proinflammatory cytokines (the so-called "cytokine storm") have been discovered through transcriptomics in viral (e.g., SARS-CoV) infections and many bacterial infections [22–25].

Transcriptional heterogeneity, within and across the complex human specimens like blood, is a major obstacle in understanding inflammation and multicellular immune response [26]. Single-cell RNA sequencing (scRNA-Seq) is a new technology that provides significantly greater benefits to researchers than bulk RNA sequencing. The reason is that individual cell gene expression is masked if the bulk tissue specimen is used for analysis [27]. Deconvolution analysis, for bulk RNA-seq signals is impossible as data for each single cell type does not exist. The advantage of single-cell sequencing is that it can overcome this issue because scRNA-Seq captures the transcriptome of individual cells; in a particular condition and/or at a particular point in time [28]. Future applications of scRNA-Seq will likely lead to major breakthroughs in systems biology of combinatorial evaluation of host gene expression and microbial transcriptome [27].

5. Recent advances of RNA sequencing in food and agriculture

The expression and biogenesis of various types of RNA molecules are analyzed by RNA-Seq in the fields of food and agriculture [29]. Differential gene expression (DGE) has been measured in plant and animal cells, throughout the cell cycle, stages of cellular development and differentiation, and in response to specific environmental factors. Also, RNA-Seq analysis has been successfully utilized to identify DEGs associated with the eggshell formation in birds, fat deposition in animals, or flavonoids, anthocyanin, etc. biosynthesis in plants [30–32]. Moreover, the role of micro-RNAs, in naturally occurring food-derived compounds, has been implicated in determining the human and microbial gene expression, which contributes to overall health and well-being of individuals [33].

Numerous databases collect RNA-Seq results of allele-specific RNA expressions, alternatively spliced or processed RNAs, and structural RNA variants that are economically important breeding traits in plants and livestock [34]. Assessment and cataloging SNP diversity in coding and non-coding regions subsequently allows to alter the path of negative or positive selection in natural populations of RNA species [35]. For example, the remodeling of root-associated transcriptomes, through the alternative RNA polyadenylation (APA), was found to increase the resistance to diverse abiotic stresses (observed in bamboo, sorghum, and arabidopsis) [36–38]. In rice species, APA site usage manipulation led to phylogenetic divergence into subspecies with beneficial agronomic traits [39].

The next few years should prove to be an exciting growth period for single-cell technology, in biomedical research [40]. Currently, there are several novel methodologies are available for generating single-cell RNA-Seq [41]. The biases that exist, within single-cell RNA-Seq protocols, represent challenges that need to be met to ensure its upward trajectory. A new biological discovery phase has just begun, and single-cell RNA-Seq has proven to be an invaluable tool, capable of guiding us through this phase.

In conclusion, we hope readers enjoy learning more about the application of RNA-Seq technologies provided throughout this book.

Acknowledgements

I am thankful for all the support received from the Association of Public Health Laboratories (APHL), during the editing of this book.

Author details

Irina Vlasova-St. Louis[1,2]

1 Department of Medicine, University of Minnesota, Minneapolis, MN, USA

2 Nevada State Public Health Laboratory, Reno, NV, USA

*Address all correspondence to: irinastl@umn.edu

References

[1] Han Y, Gao S, Muegge K, et al. Advanced applications of RNA sequencing and challenges. *Bioinform Biol Insights*; 9. Epub ahead of print 2015. DOI: 10.4137/BBI.S28991.

[2] canexiahealth. It's time for equitable access to precision oncology, https://canexiahealth.com/ (2021, accessed 26 July 2021).

[3] Zhang Y, Wang D, Peng M, et al. Single-cell RNA sequencing in cancer research. *Journal of Experimental and Clinical Cancer Research*; 40. Epub ahead of print 2021. DOI: 10.1186/s13046-021-01874-1.

[4] CarisMI. Comprehensive Tumor Profiling, https://www.carismolecularintelligence.com/ (2021, accessed 26 July 2021).

[5] www.carislifesciences.com. Where Molecular Science Meets Artificial Intelligence – Revolutionizing Cancer Care, https://www.carislifesciences.com/ (2021, accessed 26 July 2021).

[6] Westermann AJ, Vogel J. Cross-species RNA-seq for deciphering host–microbe interactions. *Nature Reviews Genetics*; 22. Epub ahead of print 2021. DOI: 10.1038/s41576-021-00326-y.

[7] Nagalakshmi U, Wang Z, Waern K, et al. The transcriptional landscape of the yeast genome defined by RNA sequencing. *Science (80-)*; 320. Epub ahead of print 2008. DOI: 10.1126/science.1158441.

[8] Rendón JM, Lang B, Llorens MR, et al. DualSeqDB: The host-pathogen dual RNA sequencing database for infection processes. *Nucleic Acids Res*; 49. Epub ahead of print 2021. DOI: 10.1093/nar/gkaa890.

[9] Lu M, Zhan X. The crucial role of multiomic approach in cancer research and clinically relevant outcomes. *EPMA Journal*; 9. Epub ahead of print 2018. DOI: 10.1007/s13167-018-0128-8.

[10] Cesur MF, Durmuş S. Systems biology modeling to study pathogen–host interactions. In: *Methods in Molecular Biology*. 2018. Epub ahead of print 2018. DOI: 10.1007/978-1-4939-7604-1_10.

[11] Moysidou CM, Owens RM. Advances in modelling the human microbiome-gut-brain axis in vitro. *Biochemical Society Transactions*; 49. Epub ahead of print 2021. DOI: 10.1042/BST20200338.

[12] Westermann AJ, Gorski SA, Vogel J. Dual RNA-seq of pathogen and host. *Nature Reviews Microbiology*; 10. Epub ahead of print 2012. DOI: 10.1038/nrmicro2852.

[13] Westermann AJ, Vogel J. Host-pathogen transcriptomics by dual RNA-seq. In: *Methods in Molecular Biology*. 2018. Epub ahead of print 2018. DOI: 10.1007/978-1-4939-7634-8_4.

[14] Oxford Genomics. Approaches to RNA sequencing, https://www.well.ox.ac.uk/ogc/approaches-to-rna-sequencing/ (accessed 28 July 2021).

[15] Viehweger A, Krautwurst S, Lamkiewicz K, et al. Direct RNA nanopore sequencing of full-length coronavirus genomes provides novel insights into structural variants and enables modification analysis. *Genome Res*; 29. Epub ahead of print 2019. DOI: 10.1101/gr.247064.118.

[16] Zhang X, Li T, Liu F, et al. Comparative Analysis of Droplet-Based Ultra-High-Throughput Single-Cell RNA-Seq Systems. *Mol Cell*; 73. Epub ahead of print 2019. DOI: 10.1016/j.molcel.2018.10.020.

[17] Huang X, Liu S, Wu L, et al. High throughput single cell RNA sequencing, Bioinformatics analysis and applications. In: *Advances in Experimental Medicine and Biology*. 2018. Epub ahead of print 2018. DOI: 10.1007/978-981-13-0502-3_4.

[18] Hwang B, Lee JH, Bang D. Single-cell RNA sequencing technologies and bioinformatics pipelines. *Experimental and Molecular Medicine*; 50. Epub ahead of print 2018. DOI: 10.1038/s12276-018-0071-8.

[19] Vlasova-St. Louis I, Chang CC, Shahid S, et al. Transcriptomic predictors of paradoxical cryptococcosis-associated immune reconstitution inflammatory syndrome. Open Forum Infect Dis 2018; 5: 1-10.

[20] Vlasova-St Louis I, Musubire AK, Meya DB, et al. Transcriptomic biomarker pathways associated with death in HIV-infected patients with cryptococcal meningitis. *BMC Med Genomics*; 14. Epub ahead of print 2021. DOI: 10.1186/s12920-021-00914-1.

[21] Mohei, Hesham; Kellampalli, Usha; Vlasova-St. Louis I, Vlasova-St. Louis I. Immune Reconstitution Disorders: Spotlight on Interferons. Int J Biomed Investig 2019; 2: 1-21.

[22] Iwalokun BA, Olalekan A, Adenipekun E, et al. Improving the Understanding of the Immunopathogenesis of Lymphopenia as a Correlate of SARS-CoV-2 Infection Risk and Disease Progression in African Patients: Protocol for a Cross-sectional Study. *JMIR Res Protoc*; 10. Epub ahead of print 2021. DOI: 10.2196/21242.

[23] Li X, Liu H, Yin H, et al. Critical roles of cytokine storm and secondary bacterial infection in acute kidney injury development in COVID-19: A multi-center retrospective cohort study. *J Med Virol*. Epub ahead of print 2021. DOI: 10.1002/jmv.27234.

[24] Jin T, Mohammad M, Pullerits R, et al. Bacteria and host interplay in staphylococcus aureus septic arthritis and sepsis. Pathogens 2021; 10: 1-25.

[25] Chen YF, Chen GY, Chang CH, et al. TRAIL encapsulated to polypeptide-crosslinked nanogel exhibits increased anti-inflammatory activities in Klebsiella pneumoniae-induced sepsis treatment. Mater Sci Eng C 2019; 102: 85-95.

[26] Luo G, Gao Q, Zhang S, et al. Probing infectious disease by single-cell RNA sequencing: Progresses and perspectives. Computational and Structural Biotechnology Journal 2020; 18: 2962-2971.

[27] Butler A, Hoffman P, Smibert P, et al. Integrating single-cell transcriptomic data across different conditions, technologies, and species. Nat Biotechnol 2018; 36: 411-420.

[28] Nguyen A, Khoo WH, Moran I, et al. Single cell RNA sequencing of rare immune cell populations. *Front Immunol*; 9. Epub ahead of print 2018. DOI: 10.3389/fimmu.2018.01553.

[29] Next generation sequencing in livestock species- A Review. *J Anim Breed Genomics*; 1. Epub ahead of print 2017. DOI: 10.12972/jabng.20170003.

[30] Khan S, Wu SB, Roberts J. RNA-sequencing analysis of shell gland shows differences in gene expression profile at two time-points of eggshell formation in laying chickens. *BMC Genomics*; 20. Epub ahead of print 2019. DOI: 10.1186/s12864-019-5460-4.

[31] Yuan S, Li R, Chen S, et al. RNA-Seq Analysis of Differential Gene Expression Responding to Different Rhizobium Strains in Soybean (Glycine max) Roots. *Front Plant Sci*; 7. Epub ahead of print 2016. DOI: 10.3389/fpls.2016.00721.

[32] Zhang Y, Jiang L, Li Y, et al. Effect of red and blue light on anthocyanin accumulation and differential gene expression in strawberry (Fragaria × ananassa). *Molecules*; 23. Epub ahead of print 2018. DOI: 10.3390/molecules23040820.

[33] Otsuka K, Yamamoto Y, Matsuoka R, et al. Maintaining good miRNAs in the body keeps the doctor away?: Perspectives on the relationship between food-derived natural products and microRNAs in relation to exosomes/extracellular vesicles. *Molecular Nutrition and Food Research*; 62. Epub ahead of print 2018. DOI: 10.1002/mnfr.201700080.

[34] Jehl F, Degalez F, Bernard M, et al. RNA-Seq Data for Reliable SNP Detection and Genotype Calling: Interest for Coding Variant Characterization and Cis-Regulation Analysis by Allele-Specific Expression in Livestock Species. *Front Genet*; 12. Epub ahead of print 2021. DOI: 10.3389/fgene.2021.655707.

[35] Rexroad C, Vallet J, Matukumalli LK, et al. Genome to phenome: Improving animal health, production, and well-being - A new USDA blueprint for animal genome research 2018-2027. *Frontiers in Genetics*; 10. Epub ahead of print 2019. DOI: 10.3389/fgene.2019.00327.

[36] Chakrabarti M, de Lorenzo L, Abdel-Ghany SE, et al. Wide-ranging transcriptome remodelling mediated by alternative polyadenylation in response to abiotic stresses in Sorghum. *Plant J*; 102. Epub ahead of print 2020. DOI: 10.1111/tpj.14671.

[37] Wang T, Wang H, Cai D, et al. Comprehensive profiling of rhizome-associated alternative splicing and alternative polyadenylation in moso bamboo (Phyllostachys edulis). *Plant J*; 91. Epub ahead of print 2017. DOI: 10.1111/tpj.13597.

[38] Cao J, Ye C, Hao G, et al. Root hair single cell type specific profiles of gene expression and alternative polyadenylation under cadmium stress. *Front Plant Sci*; 10. Epub ahead of print 2019. DOI: 10.3389/fpls.2019.00589.

[39] Zhou Q, Fu H, Yang D, et al. Differential alternative polyadenylation contributes to the developmental divergence between two rice subspecies, japonica and indica. *Plant J*; 98. Epub ahead of print 2019. DOI: 10.1111/tpj.14209.

[40] Xie Y, Jiang S, Li L, et al. Single-Cell RNA Sequencing Efficiently Predicts Transcription Factor Targets in Plants. *Front Plant Sci*; 11. Epub ahead of print 2020. DOI: 10.3389/fpls.2020.603302.

[41] Haque A, Engel J, Teichmann SA, et al. A practical guide to single-cell RNA-sequencing for biomedical research and clinical applications. *Genome Medicine*; 9. Epub ahead of print 2017. DOI: 10.1186/s13073-017-0467-4.

Chapter 2

RNA Sequencing in Potentially Malignant Disorders

Ramya Ramadoss, Rajkumar Krishnan, Lekshmy Jayan and Priyadharini Shankaran

Abstract

RNA sequencing is a molecular technique which utilizes next generation sequencing to identify and quantify ribonucleic acid (RNA) in a given sample. This technique is utilized in the detection of changes in gene expression. Potentially malignant oral disorders are one of the most troublesome lesions seen in the oral cavity which predisposes to the development of oral cancer. Though there are many methods employed in the diagnosis of these disorders, biopsy followed by histological examination is the gold standard procedure followed in the diagnosis. RNA sequencing has been receiving attention among researchers. Many studies have been conducted to analyze the application of RNA sequencing in the diagnosis of PMODs as well as in the malignant transformation to oral squamous cell carcinoma. The article attempts to summarize the progress in RNA sequencing pertaining to Potentially malignant disorders.

Keywords: RNA sequencing, Potentially Malignant Disorders, Diagnostic Markers, Molecular Diagnosis, Oral cancer

1. Introduction

RNA sequencing is a molecular technique which utilizes next generation sequencing to identify and quantify ribonucleic acid (RNA) in a given sample. This technique is utilized in the detection of changes in gene expression [1]. It also detects mutations or single nucleotide polymorphisms etc. RNA sequencing has greatly replaced cDNA microarray owing to more precise reproduction of using lanes and flow cells. Another added bonus is that this method allows de novo reconstruction of the transcriptome i.e., unknown material can be analyzed [2].

Potentially malignant oral disorders are one of the most troublesome lesions seen in the oral cavity which predisposes to the development of oral cancer. As the saying goes, "prevention is better than cure" it is better to identify and tackle the lesion in the premalignant stage rather than once cancer has developed. Sarode et al. (2014) defined OSCC prone disorders as 'It is a group of disorders of varying etiologies, usually tobacco; characterized by mutagen-associated, spontaneous or hereditary alterations or mutations in the genetic material of oral epithelial cells with or without clinical and histomorphological alterations that may lead to oral squamous cell carcinoma transformation' [3].

Though there are many methods employed in the diagnosis of these disorders, biopsy followed by histological examination is the gold standard procedure followed in the diagnosis [4]. Molecular techniques like PCR, ELISA are attempted in identifying a sensitive and specific marker. A handful of markers (salivary and serum) are attempted but none are identified to lack both specificity and sensitivity ideally required by a diagnostic marker. Microarray has emerged as a promising method as it helped in comparison and analysis of multiple samples at the same time. RNA sequencing has been receiving attention among researchers [3, 4]. Many studies have been conducted to analyze the application of RNA sequencing in the diagnosis of PMODs as well as in the malignant transformation to oral squamous cell carcinoma [5, 6].

2. RNA sequencing

RNA sequencing is the molecular technique in which the quantity as well as the sequence of RNA in a biological sample tissue of choice is determined using next generation sequencing. Here, the coding and noncoding RNAs or in short the transcriptome of the gene is analyzed. It was first described a decade ago. During the infant period of this technique, the Sanger sequencing technology was used in RNA sequencing which is greatly replaced in the present by more accurate next generation sequencing technology. RNA sequencing is found to be superior to many other techniques especially tissue microarray hybridisation [7]. The proposed advantages of the former over the later are:

- Microarray hybridisation requires use of species specific probe. It cannot be used to identify a novel or unknown sequence. RNA sequencing on the other hand can be successfully employed in the identification as well as quantification of an unknown sequence.

- Background signal or non-specific binding Is less in RNA sequencing as compared to microarray.

- Quantification of the transcriptome is more reliable in RNA sequencing as compared to microarray.

- RNA sequencing has greatly replaced cDNA microarray owing to more precise reproduction of using lanes and flow cells.

3. Steps in RNA sequencing

The procedure of RNA sequencing is performed by three steps:

1. RNA isolation

2. RNA selection or depletion

3. cDNA synthesis

4. Preparing sequencing library

5. Next generation sequencing

The first step in RNA sequencing is the isolation of RNA from the sample provided. The main requirement is that the sample should possess RNA of sufficient

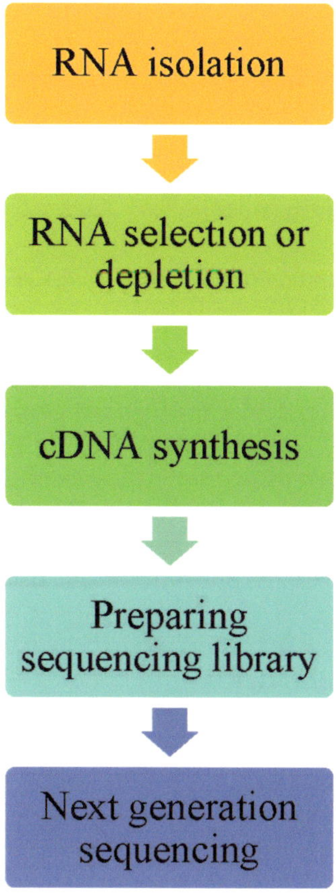

Figure 1.
Steps in RNA sequencing.

quality to enable library sequencing. Based on the quality of the RNA, measured by a bioanalyser an RNA integrity number is given ranging from 1 to 10. The number 10 shows least degradation of RNA with highest quality. RNA of low quality may result in erroneous sequencing.

The next step is the selection of RNA species from the pool of total RNA including mRNA, tRNA, rRNA etc. in this pool, around 95% is contributed by rRNA and are removed before sequencing as it may have the tendency to overshadow the read of other types of RNAs. This can be achieved by many techniques, like selecting poly A RNAs by targeted reaction with poly-T- oligos in magnetic beads. There are also commercially available kits which deplete the rRNA like Ribozero or Ribominus. In another method, the sample is treated with the enzyme, DNAse to reduce the quantity of the genetic material (DNA) in it. The quantity of RNA is determined by either capillary or gel electrophoresis. In the final step, the isolated RNA is then transcribed to DNA. DNA is more stable than RNA and thereby facilitates techniques of amplification without undergoing damage. Also, most of the sequencing libraries require DNA. This cDNA is then utilized in next generation sequencing (**Figure 1**) [8].

4. Advantages and disadvantages of RNA sequencing

Advantages of RNA sequencing are

1. We can do genome wide analysis as well as targeted analysis

2. It can analyze both novel sequences as well as known sequences

3. It has very low background noise

4. It is cheaper in comparison to Sanger sequencing

The drawback of RNA sequencing is that the depth of coverage depends on the sequenceability [9].

5. RNA sequencing in potentially malignant oral disorder

Since the scope of this current chapter is the role of RNA sequencing in the detection of potentially malignant oral disorders. This particular technique has gained attention in the identification of these disorders. The studies have focused mainly on leukoplakia, oral submucous fibrosis and also in lichen planus.

6. RNA sequencing in leukoplakia

Oral leukoplakia is the most commonly reported potentially malignant oral disorder. World Health Organization (WHO)defined oral Leukoplakia as "a white patch or plaque that cannot be characterized clinically or pathologically as any other disease" [9].

Leukoplakia is at present defined as "A white plaque of questionable risk having excluded (other) known diseases or disorders that carries no increased risk for cancer" (WHO 2005).

Diagnosis of Leukoplakia is predominantly based on clinical appearance as histological appearance seems varied. Microscopical architecture presents with a non-specific pattern of atrophy or hyperplasia. Histological evaluation is mainly done to delineate the presence and absence of epithelial dysplasia as malignant transformation rate is about 2–3% [9].

Numerous metabolic and molecular pathways are altered in leukoplakia and the disease is manifested as a culmination of all the altered metabolic pathways.

Numerous studies are conducted in analyzing the role of RNA sequencing leukoplakia. Philipone et al. in 2016 conducted a study which utilized deep RNA sequencing in the role of miRNA in both dysplastic and non-dysplastic leukoplakia. The predictive value of these markers were analyzed in both the groups. miRNA shows possible predictive value in the progression of dysplastic leukoplakia [10].

Another study by Chang et al. in 2019 conducted a study to analyze the role of potential miRNAs in the malignant transformation of leukoplakia to oral squamous cell carcinoma. They used small RNA sequencing to screening these markers in patients with leukoplakia and normal subjects. Further bioinformatics study revealed that miRNA-423-5p and miRNA-222-3p were found to have significance in the diagnosis of oral leukoplakia. RNA sequencing helped in revealing the role of these markers as potential diagnostic markers in leukoplakia as well as in the detection of malignant transformation [11].

Simming Zu et al. in 2020, analyzed the role of circular RNAs in the development of leukoplakia and identified circHLA-C has role in the progression of the disease using Sanger sequencing. They reported that levels of circHLA-C increases

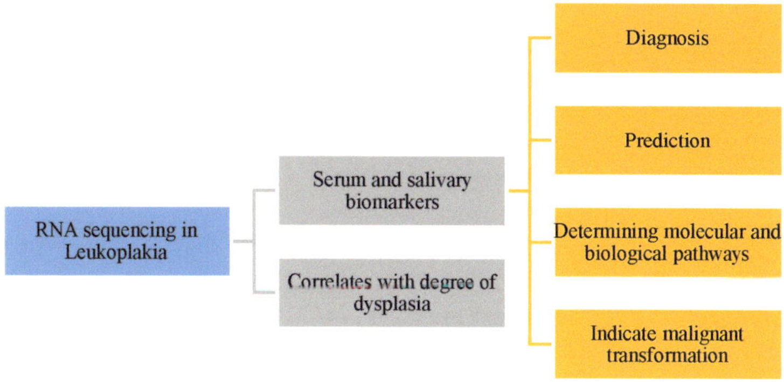

Figure 2.
Applications of RNA sequencing in leukoplakia.

with degree of dysplasia. It is a potential diagnostic marker and a genetic marker in oral leukoplakia [12].

Transcriptome analysis was conducted using RNA sequencing, differential expression in the study reported by Farah et al. (2019) which evaluated leukoplakia cases with or without dysplasia. They concluded from their study that reactive changes in the connective tissue of the lesion is an early manifestation of development of dysplasia in leukoplakia. Utilization of RNA sequencing in detection of molecular changes in oral leukoplakia will help in understanding the evasive process of development of the disease [9].

The studies conducted evaluated the role of various markers in diagnosis, prediction of the disease. The studies also revealed that the molecular pathways of the disease can be determined by RNA sequencing. It is also suggested that as the degree of dysplasia increases the progression of disease also advances. RNA sequencing may be helpful in filling the blanks in understanding the molecular and biological pathways in the development of leukoplakia (**Figure 2**).

7. RNA sequencing in oral submucous fibrosis

Oral submucous fibrosis may be defined as "an insidious, chronic disease affecting any part of the oral cavity and sometimes the pharynx. Although occasionally preceded by and/or associated with vesicle formation, it, is always associated with a juxta-epithelial inflammatory reaction followed by a fibroelastic change of the lamina propria, with epithelial atrophy leading to stiffness of the oral mucosa and causing trismus and inability to eat" [13]. 'Slowly progressive disease characterised by the fibrous bands in the oral mucosa, ultimately leading to severe restriction of mouth movement including the tongue' World Health Organization (1978).Oral submucous fibrosis is highly prevalent in South east Asia owing to the increased consumption of arecanut. Arecoline was identified as the single most important etiological agent in the development and the progression of the disease. It has a high malignant transformation rate of 7–30% [13].

Several studies analyzed the molecular profile of Oral sub mucous fibrosis and revealed that the rna profile was altered significantly in OSMF. Tsai et al. reported that role of the prime etiological agent areca nut lead to consistent elevation of miRNAs. Research evidences have substantiated the role of miRNAs in development of oral potentially malignant disorders [14].

Shangui Zhou et al. in 2019 conducted a study to determine the role of Intergenic/intertwining long RNA (IncRNAs) expression in OSMF. They used RNA

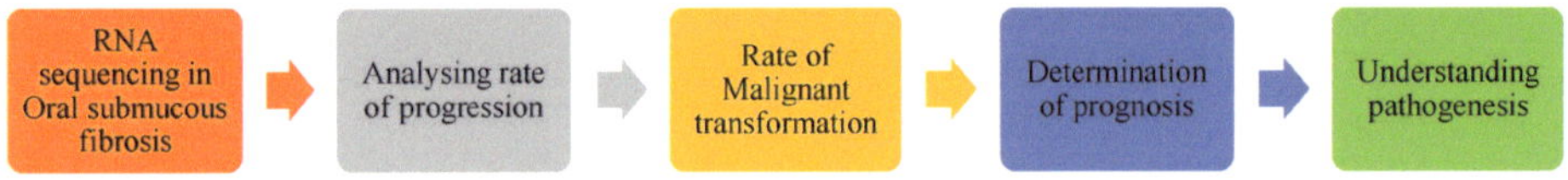

Figure 3.
Applications of RNA sequencing in oral submucous fibrosis.

sequencing to transcript the samples and found that 231 IncRNAs were upregulated and 456 were downregulated. IncRNAs were found to be associated with the regulation of progression of OSMF. These markers also plays a role in the inflammatory signaling associated with this disorder. This study was considered the first study to evaluate the role of IncRNA expression in the progression of oral submucous fibrosis [15].

Xiaohuan Zhong et al. studied the role of oral microflora in the development of oral submucous fibrosis and in the malignant transformation with continued use of arecanut. The genra of bacteria varied with site as well as conditions like alcohol or smoking. In patients with alcoholism and arecanut chewing, Prevotella was increased but at the same time Actinobacillus was reduced. But they suggested that since the sample size was small it was difficult to analyze the role of confounding factors in the oral bacterial dysbiosis [16].

The studies conducted in oral submucous fibrosis using RNA sequencing suggest the possible role in analyzing the rate of progression of the condition and also in the malignant transformation to oral squamous cell carcinoma. Also, one of the reasons for the poor prognosis of OSMF was because of lack of proper understanding of the molecular pathogenesis and the pathway of progression to oral squamous cell carcinoma (**Figure 3**).

8. RNA sequencing in oral lichen planus

Lichen planus is an inflammatory mucocutaneous disease involving skin, hair, nails and mucosal surfaces- esophageal, genital, oral, ocular, optic and less commonly bladder, nasal, laryngeal and anal mucosa. It is derived from the Greek word "*leichen*" means tree moss and Latin word "*planus*" means flat [17].

It is a T cell mediated autoimmune disorder in which cytotoxic CD8 + T cells trigger apoptosis of the basal cells of the oral epithelium. Associated with other autoimmune disorders like myasthenia gravis, alopecia, vitiligo, ulcerative colitis. The disease has been implicated to be caused by exogenous trigger also. One of the common difficulties in studying tis disorder because of the overlap between features of oral lichen planus and other oral mucosal conditions, to the highly variable application of diagnostic criteria and the potential co-existence of additional non-OLP inflammatory conditions in same patients [17].

It can be defined as "Lichen planus is a chronic immunological mucocutaneous disorder that varies in appearance from keratotic to ulcerative (Wilson)" [17].

"Oral lichen planus is a non-infectious, cytotoxic T-cell mediated, chronic inflammatory autoimmune disease affecting oral cavity, involves the oral mucosal stratified squamous epithelium and underlying lamina propria which may be accompanied by skin lesions" [17].

In a study conducted by Ku Wang et al. in 2016, the role of oral microbial flora in oral lichen planus. MiSeq sequencing was done to detect the species present in the saliva of the patients and then compared the results with that of normal patients. There was an upsurge in Porphyromonas and Solobacterium and reduced numbers of Hemophilus, Corrynebacterium, Cellulosimicrobium and Camplyobacter. In patients erosive lichen planus it was found that was a significant reduction in Streptococcus. The levels of Porphyromonas correlated with both disease progression and the immune dysregulation which is considered as the main culprit in the development of the disease [18].

Qiaozhen Yang et al. (2017) conducted a study utilizing RNA sequencing in the detection of genes responsible for malignant transformation of oral lichen planus to oral squamous cell carcinoma. Around 19 common differently expressed genes associated with oral lichen planus and OSCC were detected. Further analysis using polymerase chain reaction test revealed that among these 19 genes BCL9L, GMPS, HES1, PER2 and TSPAN33 were associated with the malignant transformation of oral lichen planus [19].

In another study conducted by Qiaozhen Yang et al. in 2017 they evaluated the role of differentially expressed genes and IncRNAs in the malignant transformation of oral lichen planus. The mapping of the IncRNAs were conducted using RNA sequencing. From the study it was concluded that keratininsation and major histocompatibility complex class I antigen processing and also the antigen presentation was activated during malignant transformation of oral lichen planus and found that numerous genes were expressed as well.

Junjun Chen et al. (2017) in their study on evaluation of the role of differentially expressed miRNAs and differentially expressed genes using next generation sequencing with DESeq. The gene expression profiling suggested a possible role in the development and progression of lichen planus [19].

In a study conducted by Keumjin Baek (2020), used high throughput sequencing of 16S rRNA gene to identify the bacterial communities present in lesions of oral lichen planus to recognize the role of these organisms in the pathogenesis of oral lichen planus. Both high throughput sequencing of 16S rRNA gene and whole genome sequencing revealed that there was an elevation in E.coli in biopsy tissues obtained from patients with oral lichen planus which is suggestive of potential role in triggering or developing the disease [20].

Studies in lichen planus done with RNA sequencing revealed newer clues as to possible role of oral microbiota in the development as well as progression of oral lichen planus. There are numerous gene expression studies which showed that there are variations in the expression profile. Like with the other two PMODs, RNA sequencing may play an important role in diagnostic and prognostic evaluation of lichen planus (**Figure 4**).

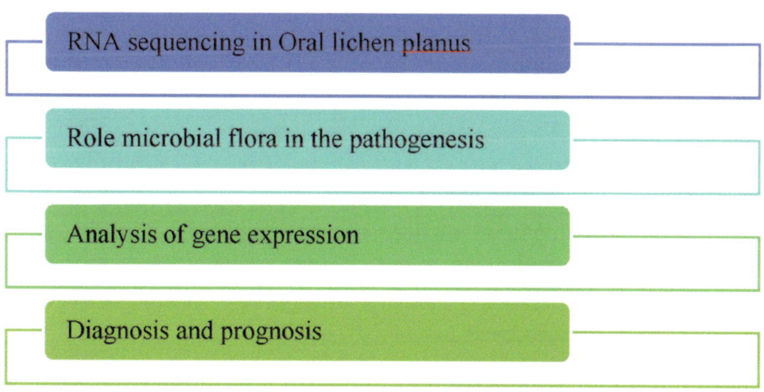

Figure 4.
Applications of RNA sequencing in oral lichen planus.

9. Scope of RNA sequencing in potentially malignant oral disorders

Potentially Malignant oral disorders have eluded the medical community for long due to lack of the right means to assess its molecular signature. Ground breaking research focusing on new molecular techniques to assess the molecular signature set by the Potentially malignant oral disorders is the need of the hour. RNA profiling serves to be a valuable tool in deciphering the molecular signature and serves as a guiding light on which the therapeutics working on similar principles can be based.

10. RNA sequencing aiding in generation of molecular signature

A significant advantage of RNA sequencing over the other diagnostic techniques is that it is based on Next generation sequencing where in the complete set of altered genome can be assessed through transcriptome analysis, thereby providing a comprehensive outlook on the genetic profile of a disease model. This elaborate guide of genetic set up of the disease serves to provide a valuable insight into the unique molecular signature of a particular disease, thereby enabling prompt and accurate diagnosis of the disease [21].

10.1 Sensitivity and specificity

RNA profiling is characteristically known for its high degree of sensitivity and specificity as it encodes for genetic alterations at the nuclear level and hence can be used as a confirmatory tool in the diagnosis of PMODs and elimination of Oral cancer in cases where clinical and histological appearance can be elusive and misleading.

10.2 RNA therapeutics

RNA therapeutics is a branch of therapeutics dealing with treatment strategies targeting the RNA profile of the disease which is unique to the individual. RNA profiling provides details about the genomic alterations unique to a particular individual. Targeted therapy towards the altered components of the genome helps eliminating the disease and offers better prognosis and avoids recurrence, thereby improving the overall survival and disease free survival rates of the individual [22].

10.3 Monitoring the prognosis and prediction of recurrence

The treatment protocol for most disorders is standard and has been in practice for decades, however, a proper protocol to assess the prognosis and the propensity for recurrence has not been established for any disease model. Obtaining the RNA profile of the individual suffering from PMODs can not only aid in diagnosis and treatment planning, but also serve as a tool in predicting the prognosis of the disease. Several genes, she upregulated serves as a poor prognostic marker where as several unregulated genes serve as markers of good prognosis and decreased recurrence rates. Hence obtaining the genomic profile through RNA sequencing can serve to be a valuable tool in predicting the same, thereby improving the overall quality of life of the individual post treatment.

10.4 RNA sequencing-a tool for research continuum

Apart from offering patient specific and disease specific outcomes, obtaining the RNA profile of a particular disease will serve as a valuable tool for furthering the cause of research pertaining to the particular disease. Most of the data obtained through RNA profiling is specific and permanent. It is not subject to change over a period of time. Thus RNA sequencing profile obtained can be used in further research, aimed at diagnosis and development of targeted therapeutics, thereby enabling continuous research up gradation.

11. Conclusion

RNA sequencing is an advanced diagnostic aid that serves to be a valuable tool in diagnosis, targeted therapy and prognostic marker for Potentially Malignant Oral Disorders. The technique is highly sensitive and specific and provides valuable information regarding the genomic set up of a particular disease. We have summarized the potential genetic alterations in PMODs and highlighted the research efforts undertaken in order to obtain the RNA profile of the particular disease. The knowledge of RNA profile added with the clinical data can serve to be a diagnostic tool par excellence in detecting Potentially Malignant Oral Disorders.

Author details

Ramya Ramadoss[1]*, Rajkumar Krishnan[2], Lekshmy Jayan[2] and Priyadharini Shankaran[2]

1 Department of Oral Pathology, Saveetha Dental College, Chennai, India

2 Department of Oral Pathology, SRM Dental College, Chennai, India

*Address all correspondence to: drramya268@gmail.com

References

[1] Kukurba KR, Montgomery SB. RNA sequencing and analysis. Cold Spring Harbor Protocols. 2015 Nov 1;2015(11):pdb-top084970.

[2] Stark R, Grzelak M, Hadfield J. RNA sequencing: the teenage years. Nature Reviews Genetics. 2019 Nov;20(11):631-656.

[3] Ganesh D, Sreenivasan P, Öhman J, Wallström M, Braz-Silva PH, Giglio D, Kjeller G, Hasseus B. Potentially malignant oral disorders and cancer transformation. Anticancer research. 2018 Jun 1;38(6):3223-3229.

[4] Ma JM, Zhou TJ, Wang R, Shan J, Wu YN, Song XL, Gu N, Fan Y. Brush biopsy with DNA-image cytometry: a useful and noninvasive method for monitoring malignant transformation of potentially malignant oral disorders. European Archives of Oto-Rhino-Laryngology. 2014 Dec;271(12): 3291-3295.

[5] Yap T, Celentano A, Seers C, McCullough MJ, Farah CS. Molecular diagnostics in oral cancer and oral potentially malignant disorders—A clinician's guide. Journal of Oral Pathology & Medicine. 2020 Jan;49(1):1-8.

[6] George A, Sreenivasan BS, Sunil S, Varghese SS, Thomas J, Gopakumar D, Mani V. Potentially malignant disorders of oral cavity. Oral Maxillofac Pathol J. 2011 Jan 1;2(1):95-100.

[7] Ozsolak F, Milos PM. RNA sequencing: advances, challenges and opportunities. Nature reviews genetics. 2011 Feb;12(2):87-98.

[8] Griffith M, Griffith OL, Mwenifumbo J, Goya R, Morrissy AS, Morin RD, Corbett R, Tang MJ, Hou YC, Pugh TJ, Robertson G. Alternative expression analysis by RNA sequencing. Nature methods. 2010 Oct;7(10):843.

[9] Farah CS, Fox SA. Dysplastic oral leukoplakia is molecularly distinct from leukoplakia without dysplasia. Oral diseases. 2019 Oct;25(7):1715-1723.

[10] Philipone E, Yoon AJ, Wang S, Shen J, Ko YC, Sink JM, Rockafellow A, Shammay NA, Santella RM. MicroRNAs-208b-3p, 204-5p, 129-2-3p and 3065-5p as predictive markers of oral leukoplakia that progress to cancer. American journal of cancer research. 2016;6(7):1537.

[11] Chang YA, Weng SL, Yang SF, Chou CH, Huang WC, Tu SJ, Chang TH, Huang CN, Jong YJ, Huang HD. A three–microRNA signature as a potential biomarker for the early detection of oral cancer. International journal of molecular sciences. 2018 Mar;19(3):758.

[12] Xu S, Song Y, Shao Y, Zhou H. Comprehensive analysis of circular RNA in oral leukoplakia: upregulated circHLA-C as a potential biomarker for diagnosis and prognosis. Annals of Translational Medicine. 2020 Nov;8(21).

[13] Pindborg JJ, Sirsat SM. Oral submucous fibrosis. Oral Surgery, Oral Medicine, Oral Pathology. 1966 Dec 1;22(6):764-779.

[14] Tsai CH, Chou MY, Chang YC. The up-regulation of cyclooxygenase-2 expression in human buccal mucosal fibroblasts by arecoline: a possible role in the pathogenesis of oral submucous fibrosis. Journal of oral pathology & medicine. 2003 Mar;32(3):146-153.

[15] Zhou S, Zhu Y, He Z, Zhang D, Guo F, Jian X, Zhang C. Long non-coding RNA expression profile

associated with malignant progression of oral submucous fibrosis. Journal of oncology. 2019 Jul 29;2019.

[16] Zhou S, Qu X, Yu Z, Zhong L, Ruan M, Ma C, Wang M, Zhang C, Jian X. Survivin as a potential early marker in the carcinogenesis of oral submucous fibrosis. Oral Surgery, Oral Medicine, Oral Pathology, Oral Radiology, and Endodontology. 2010 Apr 1;109(4):575-581.

[17] Le Cleach L, Chosidow O. Lichen planus. New England Journal of Medicine. 2012 Feb 23;366(8):723-732.

[18] Chen J, Wang Y, Du G, Zhang W, Cao T, Shi L, Wang Y, Mi J, Tang G. Down-regulation of miRNA-27b-3p suppresses keratinocytes apoptosis in oral lichen planus. Journal of cellular and molecular medicine. 2019 Jun;23(6):4326-4337.

[19] Yang Q, Guo B, Sun H, Zhang J, Liu S, Hexige S, Yu X, Wang X. Identification of the key genes implicated in the transformation of OLP to OSCC using RNA-sequencing. Oncology reports. 2017 Apr 1;37(4):2355-2365.

[20] Baek K, Choi Y. The microbiology of oral lichen planus: Is microbial infection the cause of oral lichen planus?. Molecular oral microbiology. 2018 Feb;33(1):22-28.

[21] Burnett JC, Rossi JJ. RNA-based therapeutics: current progress and future prospects. Chemistry & biology. 2012 Jan 27;19(1):60-71.

[22] Marinov GK. On the design and prospects of direct RNA sequencing. Briefings in functional genomics. 2017 Nov 1;16(6):326-335.

Chapter 3

Insights into Oropharyngeal Microbiota, Biofilms and Associated Diseases from Metagenomics and Transcriptomic Approaches

Richa Priyadarshini, Karthik Krishnan and Rashmi Niranjan

Abstract

Oral cavity is an ecologically complex environment and hosts a diverse microbial community. Most of these organisms are commensals, however, on occasion, some have the potential to become pathogenic causing damage to the human host. Complex interactions between pathogenic bacteria, the microbiota, and the host can modify pathogen physiology and behavior. Most bacteria in the environment do not exist in free-living state but are found as complex matrix enclosed aggregates known as biofilms. There has been research interest in microbial biofilms because of their importance in industrial and biomedical settings. Bacteria respond to environmental cues to fine-tune the transition from planktonic growth to biofilm by directing gene expression changes favorable for sessile community establishment. Meta-approaches have been used to identify complex microbial associations within human oral cavity leading to important insights. Comparative gene expression analysis using deep sequencing of RNA and metagenomics studies done under varying conditions have been successfully used in understanding and identifying possible triggers of pathogenicity and biofilm formation in oral commensals.

Keywords: oral microbiome, biofilms, metagenomics, metatranscriptomics, dysbiosis

1. Introduction

Human microbiome is a collection of distinct microbial communities, which colonize the human body, including the mucosal and skin environment. They include bacteria, archaea as well as fungi, viruses and protozoa. The total number of microbial cells present in human body are as abundant as the human cells and play an important role in human health and disease. It is estimated that at any point of time there are close to 1000 unique species of bacteria present on human the body [1]. The coding potential in terms of number and diversity of genes available from microbiome colonising human niches is also considerably higher than those available through human genome alone. Early studies were focussed on identifying the composition of the microbiome across various niches to create a microbial fingerprint. This was to understand if there is a core group of microbes, which humans share. However, improvement and accessibility of experimental techniques have enabled studies investigating and understanding variation of microbiome between different people and within a person over time.

In this chapter, we will explore human microbiome in general, including overview on role in health and disease, and techniques used for studying microbial content. We also present the current knowledge of oropharyngeal microbiome and sequencing studies, metatranscriptomics in particular linking them with various diseases.

2. Microbiome in health and disease

The variations in environmental and nutrient conditions present in different sites in the human anatomy lends themselves to promotion of different communities and hence unique biomes [2]. However, within a particular body site, different people may harbor different microbial content based on a variety of different factors [3].

Every individual has distinct microbiome which is the function of their immunological interaction during early development, the dietary conditions, their life style and their current health state including use of any medication [4]. Dietary conditions have significant effect on both short-term and long -term stability of microbiome. The changes in gut microbiota has been extensively studied in relations to dietary changes [5]. Life style preference of an individual has also been shown to shape the composition of microbial content. Occupation, dwelling preference, pet ownership and even exercise has shown to contribute to uniqueness of an individual's microbiome. Use of medications, especially antibiotics has been shown to have profound effect in human gut microbiota during repeat administration [6]. Primary microbial colonization occurs during and shortly after birth due to exposure to maternal microbes followed by impact of immediate environment and diet [7]. This composition is highly dynamic in nature for the first three years of life becoming relatively stable in later years.

2.1 Microbiome in health

The human microbiota over its span of development has evolved a symbiotic association with the host providing beneficial functions. Colonization of various regions of the human body by indigenous microbiota protects the host from harmful pathogens. The resident microbiota protects the host by competing with pathogenic microbes for growth and by forming a physical barrier. Release of antimicrobial substances have also been shown to stunt the growth of other microbes resulting in protection of the host [8]. Human microbiome also constantly interacts with the host to evolve, develop and maintain important processes. The initial colonization of neonates and children by microbes is responsible for evolution of immune system affecting inflammatory homeostasis [9]. Disruption of the normal colonization process such as caesarean delivery has shown to be a risk factor for allergic diseases. The absence of seeding of neonates during vaginal delivery by maternal flora has been shown to affect the presence of healthy flora and reduction in number of anti- inflammatory microbes such as *Bacteriodetes* [10].

The gut microbiota also aids in metabolism of xenobiotics and removal of toxic compounds such as pesticides, hydorcarbons etc. [11]. The urinary tract microbiota plays a role in detoxification of filtrates in bladder [12]. Plethora of metabolic genes available through the microbial genomic cache provide humans, specific and unique metabolic pathways offering ways to increase energy and nutrient extraction by enhancing the catalog of food materials [8, 13]. The gene catalog available through gut microbiome alone is estimated to be over 100 times of the total genes present in entire humans [2]. The microorganisms in digestive tract are able to break down complex carbohydrates which are not digested by human enzymatic action [14]. Similarly, action of microbes such as *Bifidobacterium spp* results in production of Vitamin K, an important coenzyme for blood coagulation process.

2.2 Microbiome in disease

Microbiome plays an important role in the health of humans as mentioned above. However, disruption of the delicate balance of the indigenous species may result in disease condition. There have been several studies to understand the effect and causation of change in microbial content during diseased condition. Dysbiosis of human microbiota can lead to infections and progression of the infection along with treatment regimen used to modify the path can significantly affect the homeostasis. *Clostridium difficile* overgrowth is a common cause of antibiotic related gut infection leading to diarrhea. An antibiotic treatment regimen can cause changes in the balance of gut microbiota through indiscriminate action on beneficial microbes. This dysbiosis leads to proliferation of opportunistic pathogen at the expense of beneficial bacteria such as butyrogenic *Fermicutes* [15].

Dysbiosis caused by alteration in composition of microbiome due to various conditions may also triggers abnormal immune response contributing to autoimmune disease [10]. Inflammatory Bowel disease has been characterized by compromise of gastrointestinal epithelial barrier including damaged mucus layer and defective cell linkages [16–18]. Butyrate, a metabolite of dietary fiber metabolism by normal gut microbiota has been shown to improve epithelial barrier function [19]. Similar to effect of depletion of butyrogenic bacteria on *Clostridium infection*, depletion of Fermicutes in gut results in increase in pro-inflammatory cytokines and reduction of anti-inflammatory cytokines leading to autoimmune condition [20]. As mentioned earlier, the gut microbiome is able to metabolize complex carbohydrates, which the host is not capable of doing and hence increases energy yield from the food ingested. This suggests that microbial composition of the digestive tract may also be one the factors along with host physiology and lifestyle, contributing to the pathophysiology of obesity [13]. Studies on mice have shown variations in indigenous microbiota of lean mice versus obese mice with *Firmicutes* dominating in obese mice as compared to prevalence of *Bacteroidetes* in lean mice [13]. The composition of gut microbiota characterized by lower diversity and plasticity has also been associated with Type 2 diabetes. Insulin resistance may be induced by species such as *Prevotella copri* and *Bacteroides vulgatus* by modulating the serum metabolome [21].

Role of microbiome in cardiovascular disease is also an active area of study. The metabolite trimethylamine N-Oxide (TMAO) which is a product of oxidation of Trimethylamine (TMA) affects cholesterol transportation and also indirectly promotes foam cell formation and hardening of arteries in animal models [22]. The gut microbiota produces TMA by metabolising l-carnitine, choline and phosphatidylcholine containing food articles. Conversely, some of the bacterial genera have also been shown to have protective effect against atherosclerosis as determined by reduction in plaque size and cholesterol deposition [23]. Cancer is another set of conditions that has seen association with microbiota but is yet to be fully defined. Metabolic processes available through the microbes have been regarded as one of the key of malignant transformation of human cells. The dysbiosis may be caused by a variety of factors including colonization by unwanted microbes as in the case of *Helicobater pylori* and its role in gastric cancer; environmental factors including diet and antibiotics [24] and microbiomes response to immunosenescence due to aging or chronic autoimmune response leading to neoplastic transformation [25].

3. Techniques to study the human microbiome

Early studies on human microbiome were limited to identifying the composition of various niches due to limitations in techniques. Early methodologies were dependent on the ability of a researcher to grow and culture microorganism under laboratory condition. This technique has obvious drawback with respect to identification of unculturable microorganisms. Subsequently, PCR and DNA hybridization based techniques provided impetus to study of microbiome. Improvement in sequencing techniques and particularly accessibility including reduction in cost has enhanced our ability to look into microbiome from various angles. The primary question, which has been fundamental to the whole scheme of things, is what are the constituents of a microbiome? What variations are seen within a person under different conditions or variations across people under similar condition? Metagenomic strategies, which are capable of identifying all the genes available in particular niches, determine the coding potential of the microbiome. However, all these techniques may not be able to answer the question of what the microbes in a habitat are doing? Metatranscriptomics approach utilizing RNA sequencing technology along with metabolomics and metaproteomics may be able to answer such questions. Each technique have set of advantages and pitfalls. Hence, use of any technique is dependent upon nature of questions researchers are attempting to address.

3.1 16S rRNA gene profile analysis

The 16 s rRNA gene encode for the small ribosome subunit RNA in microbes. Several characteristics of this 16 s rRNA gene has made it suitable for use as genetic marker for studying bacterial phylogeny and taxonomy. The gene is highly conserved between different species of bacteria and archaea, which makes it a useful housekeeping genetic marker gene. The highly conserved region is used to create universal primers for isolation of amplicons for sequencing. Apart from highly conserved regions, the 16S rRNA also has nine-hypervariable (named V1- V9) regions scattered across the gene. Sequencing of the amplicons and mapping of the hypervariable regions to a database of known 16SrRNA sequences allow for taxonomic identification of a microbe in a sample. The sequencing of 16sRNA gene has become the mainstay of identifying and quantifying bacteria present in a sample. However, the use of 16S rRNA gene sequencing does have certain limitations, which has to be taken into account. Some bacteria have multiple copies for the gene arranged as gene family or operons, which may introduce bias [26] with the analysis. Bias may also be introduced by PCR primer favoring specific group or selection of specific hypervariable region [27]. The reduction in cost of sequencing after introduction of NGS technologies and simplicity of use of 16S rRNA as genetic marker made a significant impact on studying microbiome. However, inability in identification of species or strain level resolution by use of 16S rRNA technique is a limiting factor in its wider use.

3.2 Metagenomic analysis

Metagenomic process involves isolation of total DNA from microbiome sample, which is then fragmented into smaller pieces. The adapters are ligated to 3′ and

5′ repaired ends of the DNA library followed by amplification and sequencing. One of the major problems in a human microbiome project is contamination of human DNA with the microbiome sample which can in some cases be upto 99% of total DNA [28]. Hence, for higher coverage, a large number of sequence reads are required for obtaining reasonable results pertaining to microbiome which in turn increases the cost. In contrast, 16S rRNA profiling requires little amount of DNA. Metagenomics approach allow us to understand the genetic potential available within the microbiome for various metabolic processes which is not possible with 16 s rRNA method. Metagenomic technique can be used in variety of different ways which are tremendously useful for identifying novel metabolic pathways, enzymatic functions etc. The tremendous genetic potential locked in unculturable microbes can be teased out by metagenomics approached. The metagenomic gene sequence identified for specific gene of interest can be further cloned and expressed.

3.3 Metatranscriptomic analysis

A Metatranscriptomics experiment is similar to metagenomics in its approach, where the total RNA is isolated from a microbiome sample followed by fragmentation and cDNA synthesis. Again, the 3′ and 5′ ends of the DNA are repaired and ligated with adapter before sequencing. The biases introduced due to use of amplification step during cDNA synthesis may affect exact quantification sometime [29]. The sequence reads can be mapped to reference genome/gene or used to assemble the transcriptome *de novo.*

4. Oral cavity and microbial niches

The oral cavity has large number of surfaces and environment for development of distinct niches. The variable environmental conditions like changes in oxygen concentration, variability in nutrients availability, physical interventions liking brushing of teeth and presence of saliva affecting the pH ranges; all contribute to growth of organisms creating distinct niches. Studies done on different microbial communities in oral cavity have found consistent similarities in composition, which were clearly distinct from microbiomes found in other parts of human body. However, there are variability in proportions of the organisms present [30]. The plethora of physical surfaces available provides opportunity for development of distinct biofilm communities.

5. Biofilms in oral cavity

A surface associated community of microbial cells is termed as biofilm, the association being irreversible in nature. Monospecies biofilms are rarely found in natural conditions. Van Leeuwenhoek was the first to observe microorganisms on tooth surfaces by the use of his own microscope [31, 32] leading to revelation of existence of microbial cells as complex- structured interspecies communities in nature. In biofilm, the microbial cells are enclosed in an extracellular polymeric substances matrix (EPS) which is primarily composed of polysaccharides. This EPS accounts for 50–90% of dry biomass of biofilm [33, 34]. Biofilm-associated cells differ from their planktonic counterparts in extracellular polymeric substance (EPS) matrix formation, reduced growth rates, and the up- and down- regulation of specific genes [35, 36]. Biofilm has a defined three dimensional structure attached to a surface. The surface to which these cells adhere can be any solid surface exposed to aqueous environments, in human body it is especially on mucous membranes and other surfaces such as on indwelling catheters, ports, implants, artificial heart valves, endotracheal tubes and prosthetic joints [32, 37, 38].

One such dwelling of biofilm is the oral cavity of humans, identified as the second most diverse and complex microbiome after colon. Oral cavity provides many different surfaces to the microbiota to attach to such as tooth enamel, and mucous membranes lining tongue, gum, hard-soft palate and cheek [39, 40]. The different characteristic properties of these surfaces contribute to complex and diverse populations in oral cavity. Biofilm in the form of supragingival and subgingival plaque is the etiologic agent in dental caries and periodontal diseases [41–43]. The physical and chemical properties of EPS vary based on synthesizing organism and environment of growth.

5.1 Biofilm formation stages

Oral microbiota is the major causative agent of dental caries and periodontitis, two most prevalent diseases in developing and developed countries altogether. Oral biofilms have been commonly termed as "plaque". Oral biofilms are dynamic in nature both spatially and temporally [44]. The formation of oral biofilm is a complex process occurring in stages: (a) reversible adhesion to the surface, (b) EPS production and irreversible adherence, (c) biofilm maturation, (d) biofilm dispersion and recolonization [45].

The initial step i.e., irreversible adhesion of bacterial cells to the substrate surface is the most crucial stage for biofilm formation. After the completion of first step of initial attachment bacterial life cycle can proceed to one of the two pathways: biofilm formation or planktonic phase, depending on environmental conditions [46, 47].

5.1.1 Reversible association

Pellicle formation is the first requirement for formation of oral biofilm. Pellicle formation occurs as soon as tooth surfaces are cleaned and exposed to moist oral cavity favouring attachment of microbiota [48]. Thin acquired pellicle predominantly comprises of saliva glycoproteins, such as proline-rich proteins, α-amylase, mucins, and agglutinin [49]. The predominant initial colonizers of teeth are Gram-positive facultative anaerobic cocci and rods, especially of *Streptococcus* and *Actinomyces* species [50]. Pellicle formation is followed by secretion of EPS and biofilm development.

5.1.2 EPS production and irreversible adhesion

Immediately after attachment of early colonizers to the pellicle, bacteria begins to secrete EPS laying the foundation for biofilm maturation [51]. Mechanism of secretion of EPS varies with Gram positive and Gram negative bacteria. Gram-positive oral bacteria synthesizes EPS via glucosyltransferases gene. This family of Gtf gene uses sucrose as substrate to synthesize soluble and insoluble glucans. Though GtfB, GtfC, and GtfD, produced by *Streptococcus mutans* have been well characterized but structural confirmation of only GtfC is available, therefore, the mechanism of EPS secretion is not well understood [52–54]. Oral microbiota is rich in non-Gtf-synthesizing microbes too such as *Lactobacillus casei*, and *Candida albicans* which do not produce glucans until and unless bound by *S. mutans* Gtfs [55].

5.1.3 Biofilm maturation

EPS is the scaffold holding all the oral microbes together, where growth of bacteria takes place. After EPS formation, different oral bacteria come and adhere to already adhered pioneer microbes. Different species of bacteria coaggregate using unique mechanisms of recognizing polysaccharides or protein receptors present on the early colonizers by late colonizers [51, 56, 57]. With time this leads to fully structural and functional complex biofilm. Though bacteria coaggregate with each other in biofilm formation but this process is species specific. Previous studies have shown that *S. mutans* aggregates with *Fusobacterium nucleatum* but not with *Porphyromonas gingivalis*. This is because one bacterial cell has several receptors complementary to adhesions present on other bacterial cell and if two bacterial cells recognize the same receptor, the two cells would compete for the same binding site [58, 59]. The complex structural association of bacteria with different receptors recognizable by adhesions of different bacterial species is known as coaggregation bridges, one of the most crucial requirement for biofilm growth and maturation. In oral cavity *F. nucleatum* is one of the best known coaggregation bridge species [60]. The components of mature biofilm differ from the initial biofilm components.

5.1.4 Biofilm dispersion and recolonization

Dispersion and recolonization is the final stage of biofilm development. It is a complex process involving environmental signals, transduction pathways, effector molecules and their response [61]. Bacterial biofilm dispersal is divided into distinguishing phases: (i) detachment of cells, (ii) translocation of the cells to a new location, and (iii) cell adhesion to a substrate in the new location [62]. Biofilm dispersal mechanism can be divided into two broad categories: active and passive. The mechanism initiated by the bacteria themselves comes under active category whereas those that are the result of external forces like abrasion or human intervention belong to passive dispersal [63]. During active dispersion, the bacteria itself initiates mechanisms in response to a trigger, mostly change in the environment of oral cavity, which is felt by the bacteria thus inducing the release of cells from the biofilm [64].

5.2 Components of oral biofilm

Most of the biofilm matrix comprises of water. The other components of biofilm are EPS matrix, microbes, DNA, RNA and proteins.

5.2.1 Exopolysaccharides (EPS)

Exopolysaccharides (EPS) are the major components of biofilm produced by the bacteria in the biofilm; in fact, they can be designated as the backbone of biofilms. Composition of EPS varies a lot. Exopolysaccharides synthesized by microbes are mostly polyanionic because of presence of uronic acids, ketal-linked pyruvate and inorganic residues, such as phosphate [65], although a few EPS such as of Staphylococcus might be polycationic and some are neutral [66]. Many bacterial EPS possess structural sequences of 1,3- or 1,4-β-linked hexose residues [67, 68], which provides rigidity to biofilm. The major EPS matrix components in oral biofilms are polysaccharides, particularly glucans and fructans produced by oral microbiota.

5.2.2 Microorganism involved in oral biofilm formation

Streptococcus mutans, Streptococcus sanguis, Streptococcus oralis and Streptococcus gordonii, Streptococcus mitis, Streptococcus infantis, Streptococcus parasanguinis, Streptococcus cristatus and Streptococcus bovis are the major oral biofilm forming bacteria. Though *Streptococcus* is the dominant species in oral biofilm but *Veillonella, Gemella, Prevotella, Niesseria, Actinomyces, Haemophilus, Propionibacterium, Capnocytophaga, Eikenella,* and *Rothia* are also found. All these species fall under the category of early colonizers [45, 69]. *Eubacterium, Treponema, Porphyromonas gingivalis, Aggregatibacter actinomycetemcomitans, Fusobacterium nucleatum, and Prevotella intermedia* are among the late colonizers of oral biofilm. Microbial composition of oral biofilms varies with its stage; early colonizers give way to late colonizers. Among *Streptococcus* species, *S. vestibularis* makes 40% of the total biofilm microbes [70]. *Streptococcus mutans* aggregates with *Candida albicans* which in turn coaggregates with other *Streptococcus* species causing formation of multilayered biofilm structure [71, 72].

5.2.3 Extracellular DNA (eDNA)

eDNA is another major constituent of oral biofilms. Since DNA is a very stable molecule, it survives several years. This DNA is called extracellular DNA (eDNA) [73]. Many studies have confirmed the presence of eDNA in biofilm matrix. One such evidence is electron microscopic images of dental plaques, which showed it to be rich in membrane vesicles a reservoir of eDNA [74]. Even though eDNA has been identified in many monospecies biofilm model but little to no knowledge is available about its role in mixed-species biofilms. Cell lysis is one of the major mechanism responsible for eDNA release in biofilm matrix [75]. This cell lysis could be either by antimicrobial agents or by bacteriocins. Secretions of vesicles and viral particles are another source of eDNA in biofilms [76]. eDNA performs some very important functions in biofilms such as adhesion in biofilm structure, protection against antimicrobial agents, genetic exchange in biofilm and nutrient storage [77, 78]. Strong evidence supports adhesion nature of eDNA as seen in *Enterococcus faecalis* where eDNA enhances the adhesion of *E. faecalis* cells in periodontic infections [79]. Second function of eDNA is protection against antimicrobial agents.

5.3 Metatranscriptomics of oral biofilm assembly and maturation

The complexities of oral niche, which results in constant changes in environmental conditions, has always interested researchers. The formation of oral biofilm has been studied on compositional level mainly attempting to identify the key players during health and disease. A significant study by Edlund et al. [80] showed the power of metatransciptomic approach by attempting to dissect the oral biofilm assembly and maturation process. The group created a simulated environment for growth of oral plaque biofilm by seeding culture with saliva samples from healthy individuals. Biofilm samples were collected for analysis at various times for pH and sequencing. In this way, the researchers were attempting understand the changes in expression of genes at the community level over time. They saw a drop in pH level from 5.5 to 4.7 at 6 to 9 hr. shift. Several members like *Streptococcus parasanguinis, S. vestibularis, S. salivarius, Veillonell* and *Lactobacillus fermentum* genome showed increased gene activity during shift to lower pH conditions. *Granulicatella adiacens, G. elegans, L. salivarius and Streptococcus pneumonia* showed significant downregulation in their gene activity. Shift in overall community functions were detected during maturation process. Increase in gene expression of L and D lactate dehyrogenases were seem during shift to lower pH. *L. fermentum*, *Veillonella sp*. and *Streptococcus sp* like *S. mitis* were upregulating the lactate metabolism genes. Also, increase in expression of hydrogen peroxide detoxification genes were observed which were driven by *Streptococcus* and *Veillonell*a sp. members.

6. Oral microbiome and diseases

6.1 Oral diseases

Dental caries has been shown to be caused by acidogenic and aciduric bacterial species such as *Strepotococcus mutans* and *Lactobacillus sp* [81]. Metatranscriptomic study done on active caries samples showed upto 400 metabolically active bacterial species with members of genera *Streptococcus* and *Veillonella* dominating [82]. Community-wide expression profile of caries sample showed gene activity associated with oxidative stress, superoxide and peroxide detoxification [83]. Metatranscriptomic studies on periodontal disease samples showed high level of functional conservation even though there were variation in composition of microbes [84]. These studies suggested that instead of specific pathogens, some disease conditions have to be looked at from the perspective of community function. The studies have shown several metabolic processes related to flagellar motility, peptide transfer and iron acquisition overrepresented. Metatranscriptomic studies on the 'red complex' consisting of *Porphyromonas gingivalis, Treponema denticola* and *Tanerella forsythia* considered the primary periodontal pathogens showed high expression of metalloproteases, motility related genes, peptidases and iron metabolism genes.

6.2 Non-oral diseases

Oral cavity is not an isolated niche and has connections to several parts of the body. This connection exposes other areas to oral microbiome and in case of dysbiosis of the microbial composition, possible disease condition. Poor oral hygiene resulting in tooth loss and periodontal diseases has been shown to have a significant association with respiratory tract infections [85], cystic fibrosis [86] and Chronic Obstructive Pulmonary Disease (COPD) [87]. Displacement of benign residents like *Prevotella spp*. and *Veillonella spp*. by pathogens like *Pseudomonas aeruginosa* and *Klebsiella pneumonia* has been shown to be one of the factors linked to ICU stay associated respiratory tract infection [88, 89]. In case of Cystic Fibrosis, the oral cavity has been proposed to be a potential reservoir for *Pseudomonad aeruginosa* [86]. *P. aeruginosa* is one of the chronic colonizer associated with Cystic Fibrosis. Metabolites produced by oral microbes like 2, 3 butanedione gas possibly produced by *Streptococcus spp*. acts as substrate for phenazines production by *P. aeruginosa* in CF lung [90].

Infectious agents and chronic infections caused by them has been shown to be linked with atleast 13% of global cancer burden [91]. Periodontitis and resulting dysbiosis in oral microbiome has been linked to variety to cancer pathologies including but not limited to oral, esophageal, colorectal, gastric and pancreatic cancers [92]. Several hypotheses have been suggested to explain this association; production of metabolites which may act as carcinogen [93], increase in inflammatory immune response [94], and increase in cancer-linked virus burden [95]. NGS- based study have shown association of genera *Lactobacillus* and *Rothia* with colorectal cancers [96] Similarly, keystone pathogens like *Porphyomonas gingivalis*

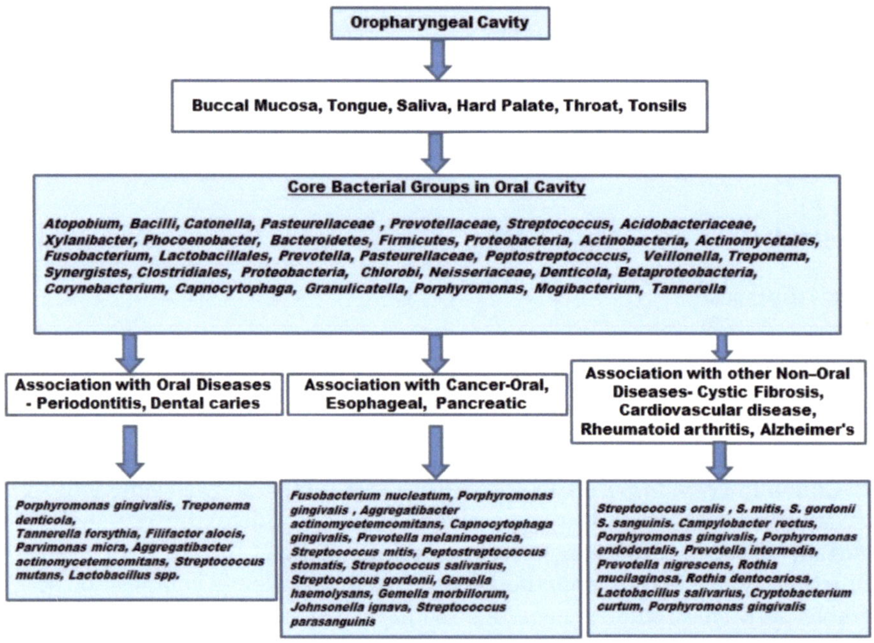

Figure 1.
A schematic representation of microbiome content of oropharyngeal cavity [100] and association under various diseased conditions [101].

and *Aggregatibacter actinomycetemcomitans* has been shown to be abundant in pancreatic cancer samples [97]. The systemic inflammation caused by periodontitis is also been linked to cardiovascular diseases [98] and diabetes [99] (**Figure 1**).

7. Conclusion

Improvements in experimental techniques have significantly enhanced the ability of researchers to expand the study of microbiome and understand its function in the context of human health. Current metagenomics studies of oral microbiome has given an opportunity to make an informed assumption regarding structure of oral microbiome and association with diseased conditions. However, functional level characterization of components of oral flora with respect to host interaction and disease condition during dysbiosis is still lacking. Maturation of transcriptomics, proteomics and metabolomics approaches when used in combination provides an exciting opportunity for functional analysis of interaction between host and microbiome.

Acknowledgements

The RP research group is supported by Shiv Nadar University, DST SERB Young Scientist grant and CSIR-EMR grant.

Conflict of interest

The authors declare no conflict of interest.

Author details

Richa Priyadarshini*, Karthik Krishnan* and Rashmi Niranjan
Department of Life Sciences, School Natural Sciences, Shiv Nadar University, India

*Address all correspondence to: richa.priyadarshini@snu.edu.in
and karthik.krishnan@snu.edu.in

References

[1] Turnbaugh PJ, Ley RE, Hamady M, Fraser-Liggett CM, Knight R, Gordon JI. The human microbiome project. Nature. 2007;449(7164):804-810.

[2] Ursell LK, Metcalf JL, Parfrey LW, Knight R. Defining the human microbiome. Nutrition reviews. 2012;70 Suppl 1:S38–S44.

[3] Human Microbiome Project C. Structure, function and diversity of the healthy human microbiome. Nature. 2012;486(7402):207-214.

[4] Integrative HMPRNC. The Integrative Human Microbiome Project: dynamic analysis of microbiome-host omics profiles during periods of human health and disease. Cell host & microbe. 2014;16(3):276-289.

[5] Albenberg LG, Wu GD. Diet and the intestinal microbiome: associations, functions, and implications for health and disease. Gastroenterology. 2014;146(6):1564-1572.

[6] Dethlefsen L, Relman DA. Incomplete recovery and individualized responses of the human distal gut microbiota to repeated antibiotic perturbation. Proceedings of the National Academy of Sciences of the United States of America. 2011;108 Suppl 1:4554-4561.

[7] Koenig JE, Spor A, Scalfone N, Fricker AD, Stombaugh J, Knight R, et al. Succession of microbial consortia in the developing infant gut microbiome. Proceedings of the National Academy of Sciences of the United States of America. 2011;108 Suppl 1:4578-4585.

[8] Gill SR, Pop M, Deboy RT, Eckburg PB, Turnbaugh PJ, Samuel BS, et al. Metagenomic analysis of the human distal gut microbiome. Science. 2006;312(5778):1355-1359.

[9] Cingi C, Bayar Muluk N, Scadding GK. Will every child have allergic rhinitis soon? International journal of pediatric otorhinolaryngology. 2019;118:53-58.

[10] Ipci K, Altintoprak N, Muluk NB, Senturk M, Cingi C. The possible mechanisms of the human microbiome in allergic diseases. European archives of oto-rhino-laryngology : official journal of the European Federation of Oto-Rhino-Laryngological Societies. 2017;274(2):617-626.

[11] Ursell LK, Knight R. Xenobiotics and the human gut microbiome: metatranscriptomics reveal the active players. Cell metabolism. 2013;17(3):317-318.

[12] Thomas-White K, Brady M, Wolfe AJ, Mueller ER. The bladder is not sterile: History and current discoveries on the urinary microbiome. Current bladder dysfunction reports. 2016;11(1):18-24.

[13] Turnbaugh PJ, Ley RE, Mahowald MA, Magrini V, Mardis ER, Gordon JI. An obesity-associated gut microbiome with increased capacity for energy harvest. Nature. 2006;444(7122):1027-1031.

[14] Goodrich JK, Davenport ER, Waters JL, Clark AG, Ley RE. Cross-species comparisons of host genetic associations with the microbiome. Science. 2016;352(6285):532-535.

[15] Antharam VC, Li EC, Ishmael A, Sharma A, Mai V, Rand KH, et al. Intestinal dysbiosis and depletion of butyrogenic bacteria in Clostridium difficile infection and nosocomial diarrhea. Journal of clinical microbiology. 2013;51(9):2884-2892.

[16] Klag T, Stange EF, Wehkamp J. Defective antibacterial barrier in

inflammatory bowel disease. Digestive diseases. 2013;31(3-4):310-316.

[17] Atreya R, Neurath MF. IBD pathogenesis in 2014: Molecular pathways controlling barrier function in IBD. Nature reviews Gastroenterology & hepatology. 2015;12(2):67-68.

[18] Lee SH. Intestinal permeability regulation by tight junction: implication on inflammatory bowel diseases. Intestinal research. 2015;13(1):11-18.

[19] Peng L, Li ZR, Green RS, Holzman IR, Lin J. Butyrate enhances the intestinal barrier by facilitating tight junction assembly via activation of AMP-activated protein kinase in Caco-2 cell monolayers. The Journal of nutrition. 2009;139(9):1619-1625.

[20] Bejaoui M, Sokol H, Marteau P. Targeting the Microbiome in Inflammatory Bowel Disease: Critical Evaluation of Current Concepts and Moving to New Horizons. Digestive diseases. 2015;33 Suppl 1:105-112.

[21] Pedersen HK, Gudmundsdottir V, Nielsen HB, Hyotylainen T, Nielsen T, Jensen BA, et al. Human gut microbes impact host serum metabolome and insulin sensitivity. Nature. 2016;535(7612):376-381.

[22] Troseid M. Gut microbiota and acute coronary syndromes: ready for use in the emergency room? European heart journal. 2017;38(11):825-827.

[23] Chan YK, Brar MS, Kirjavainen PV, Chen Y, Peng J, Li D, et al. High fat diet induced atherosclerosis is accompanied with low colonic bacterial diversity and altered abundances that correlates with plaque size, plasma A-FABP and cholesterol: a pilot study of high fat diet and its intervention with Lactobacillus rhamnosus GG (LGG) or telmisartan in ApoE(−/−) mice. BMC microbiology. 2016;16(1):264.

[24] Zitvogel L, Galluzzi L, Viaud S, Vetizou M, Daillere R, Merad M, et al. Cancer and the gut microbiota: an unexpected link. Science translational medicine. 2015;7(271):271ps1.

[25] Kamada N, Seo SU, Chen GY, Nunez G. Role of the gut microbiota in immunity and inflammatory disease. Nature reviews Immunology. 2013;13(5):321-335.

[26] Kembel SW, Wu M, Eisen JA, Green JL. Incorporating 16S gene copy number information improves estimates of microbial diversity and abundance. PLoS computational biology. 2012;8(10):e1002743.

[27] Bikel S, Valdez-Lara A, Cornejo-Granados F, Rico K, Canizales-Quinteros S, Soberon X, et al. Combining metagenomics, metatranscriptomics and viromics to explore novel microbial interactions: towards a systems-level understanding of human microbiome. Computational and structural biotechnology journal. 2015;13:390-401.

[28] Breitenstein S, Tummler B, Romling U. Pulsed field gel electrophoresis of bacterial DNA isolated directly from patients' sputa. Nucleic acids research. 1995;23(4):722-723.

[29] Liu D, Graber JH. Quantitative comparison of EST libraries requires compensation for systematic biases in cDNA generation. BMC bioinformatics. 2006;7:77.

[30] Costello EK, Lauber CL, Hamady M, Fierer N, Gordon JI, Knight R. Bacterial community variation in human body habitats across space and time. Science. 2009;326(5960):1694-1697.

[31] Gest H. The discovery of microorganisms by Robert Hooke and Antoni Van Leeuwenhoek, fellows of the Royal Society. Notes and records

of the Royal Society of London. 2004;58(2):187-201.

[32] Donlan RM. Biofilms: microbial life on surfaces. Emerg Infect Dis. 2002;8(9):881-890.

[33] Flemming HC, Wingender J. The biofilm matrix. Nat Rev Microbiol. 2010;8(9):623-633.

[34] Di Martino P. Extracellular polymeric substances, a key element in understanding biofilm phenotype. AIMS Microbiol. 2018;4(2):274-288.

[35] Gebreyohannes G, Nyerere A, Bii C, Sbhatu DB. Challenges of intervention, treatment, and antibiotic resistance of biofilm-forming microorganisms. Heliyon. 2019;5(8):e02192.

[36] Sharma D, Misba L, Khan AU. Antibiotics versus biofilm: an emerging battleground in microbial communities. Antimicrob Resist Infect Control. 2019;8:76.

[37] Donlan RM. Biofilm formation: a clinically relevant microbiological process. Clin Infect Dis. 2001;33(8):1387-1392.

[38] Nandakumar V, Chittaranjan S, Kurian VM, Doble M. Characteristics of bacterial biofilm associated with implant material in clinical practice. Polymer journal. 2013;45(2):137-152.

[39] Deo PN, Deshmukh R. Oral microbiome:Unveilingthefundamentals. Journal of oral and maxillofacial pathology: JOMFP. 2019;23(1):122.

[40] Kilian M, Chapple I, Hannig M, Marsh P, Meuric V, Pedersen A, et al. The oral microbiome–an update for oral healthcare professionals. British dental journal. 2016;221(10):657-666.

[41] Lazar V, Ditu L-M, Curutiu C, Gheorghe I, Holban A, Popa M, et al. Impact of dental plaque biofilms in periodontal disease: Management and future therapy. Periodontitis: A Useful Reference; Arjunan, P, Ed; InTech Open: London, UK. 2017:11-42.

[42] Loesche WJ. Microbiology of dental decay and periodontal disease. Medical Microbiology 4th edition. 1996.

[43] Biradar B, Biradar S, Malhan B, Arvind M, Arora M. Oral biofilm-a review. International Journal of Oral Health Dentistry. 2017;3(3):142-148.

[44] Scannapieco FA. The oral microbiome: its role in health and in oral and systemic infections. Clinical Microbiology Newsletter. 2013;35(20):163-169.

[45] Krzyściak W, Jurczak A, Piątkowski J. The role of human oral microbiome in dental biofilm formation. Microbial Biofilms—Importance and Applications InTech. 2016:329-382.

[46] Donlan RM, Costerton JW. Biofilms: survival mechanisms of clinically relevant microorganisms. Clin Microbiol Rev. 2002;15(2):167-193.

[47] Kostakioti M, Hadjifrangiskou M, Hultgren SJ. Bacterial biofilms: development, dispersal, and therapeutic strategies in the dawn of the postantibiotic era. Cold Spring Harbor perspectives in medicine. 2013;3(4):a010306.

[48] Marsh PD, editor Dental plaque as a biofilm and a microbial community–implications for health and disease. BMC Oral health; 2006: BioMed Central.

[49] Heller D, Helmerhorst EJ, Oppenheim FG. Saliva and Serum Protein Exchange at the Tooth Enamel Surface. Journal of dental research. 2017;96(4):437-443.

[50] Larsen T, Fiehn NE. Dental biofilm infections–an update. Apmis. 2017;125(4):376-384.

[51] Huang R, Li M, Gregory RL. Bacterial interactions in dental biofilm. Virulence. 2011;2(5):435-444.

[52] Cugini C, Shanmugam M, Landge N, Ramasubbu N. The role of exopolysaccharides in oral biofilms. Journal of dental research. 2019;98(7):739-745.

[53] Lemos J, Palmer S, Zeng L, Wen Z, Kajfasz J, Freires I, et al. The biology of Streptococcus mutans. Gram-Positive Pathogens. 2019:435-448.

[54] Matsumoto-Nakano M. Role of Streptococcus mutans surface proteins for biofilm formation. Japanese Dental Science Review. 2018;54(1):22-29.

[55] Bowen W, Koo H. Biology of Streptococcus mutans-derived glucosyltransferases: role in extracellular matrix formation of cariogenic biofilms. Caries research. 2011;45(1):69-86.

[56] Kolenbrander PE, Andersen RN, Blehert DS, Egland PG, Foster JS, Palmer RJ. Communication among oral bacteria. Microbiology and molecular biology reviews. 2002;66(3):486-505.

[57] Kolenbrander PE, London J. Adhere today, here tomorrow: oral bacterial adherence. Journal of bacteriology. 1993;175(11):3247.

[58] Kolenbrander PE, Palmer RJ, Periasamy S, Jakubovics NS. Oral multispecies biofilm development and the key role of cell–cell distance. Nature Reviews Microbiology. 2010;8(7):471-480.

[59] Rickard AH, Gilbert P, High NJ, Kolenbrander PE, Handley PS. Bacterial coaggregation: an integral process in the development of multi-species biofilms. Trends in microbiology. 2003;11(2):94-100.

[60] Bradshaw DJ, Marsh PD, Watson GK, Allison C. Role of Fusobacterium nucleatum and coaggregation in anaerobe survival in planktonic and biofilm oral microbial communities during aeration. Infection and immunity. 1998;66(10):4729-4732.

[61] Karatan E, Watnick P. Signals, regulatory networks, and materials that build and break bacterial biofilms. Microbiology and molecular biology reviews : MMBR. 2009;73(2):310-347.

[62] Kaplan Já. Biofilm dispersal: mechanisms, clinical implications, and potential therapeutic uses. Journal of dental research. 2010;89(3):205-218.

[63] Lawrence JR, Scharf B, Packroff G, Neu TR. Microscale evaluation of the effects of grazing by invertebrates with contrasting feeding modes on river biofilm architecture and composition. Microbial ecology. 2002;44(3):199-207.

[64] Bowen WH, Burne RA, Wu H, Koo H. Oral biofilms: pathogens, matrix, and polymicrobial interactions in microenvironments. Trends in microbiology. 2018;26(3):229-242.

[65] Shatwell KP, Sutherland IW, Ross-Murphy SB. Influence of acetyl and pyruvate substituents on the solution properties of xanthan polysaccharide. International journal of biological macromolecules. 1990;12(2):71-78.

[66] Lembre P, Lorentz C, Di Martino P. Exopolysaccharides of the biofilm matrix: a complex biophysical world. The complex world of polysaccharides. 2012:371-392.

[67] Sutherland IW. Biofilm exopolysaccharides: a strong and sticky framework. Microbiology. 2001;147(1):3-9.

[68] Vu B, Chen M, Crawford RJ, Ivanova EP. Bacterial extracellular polysaccharides involved in biofilm formation. Molecules. 2009;14(7):2535-2554.

[69] Kolenbrander PE, Egland PG, Diaz PI, Palmer RJ, Jr. Genome-genome interactions: bacterial communities in initial dental plaque. Trends Microbiol. 2005;13(1):11-15.

[70] Edlund A, Yang Y, Hall AP, Guo L, Lux R, He X, et al. An in vitro biofilm model system maintaining a highly reproducible species and metabolic diversity approaching that of the human oral microbiome. Microbiome. 2013;1(1):1-17.

[71] Kim D, Sengupta A, Niepa TH, Lee B-H, Weljie A, Freitas-Blanco VS, et al. Candida albicans stimulates Streptococcus mutans microcolony development via cross-kingdom biofilm-derived metabolites. Scientific Reports. 2017;7(1):1-14.

[72] Khoury ZH, Vila T, Puthran TR, Sultan AS, Montelongo-Jauregui D, Melo MAS, et al. The role of Candida albicans secreted polysaccharides in augmenting Streptococcus mutans adherence and mixed biofilm formation: In vitro and in vivo studies. Frontiers in microbiology. 2020;11:307.

[73] Nicholls H. Ancient DNA comes of age. PLoS biology. 2005;3(2):e56.

[74] Holliday R, Preshaw PM, Bowen L, Jakubovics NS. The ultrastructure of subgingival dental plaque, revealed by high-resolution field emission scanning electron microscopy. BDJ open. 2015;1:15003.

[75] Jakubovics N, Shields R, Rajarajan N, Burgess J. Life after death: the critical role of extracellular DNA in microbial biofilms. Letters in applied microbiology. 2013;57(6):467-475.

[76] Ibáñez de Aldecoa AL, Zafra O, González-Pastor JE. Mechanisms and regulation of extracellular DNA release and its biological roles in microbial communities. Frontiers in microbiology. 2017;8:1390.

[77] Okshevsky M, Meyer RL. The role of extracellular DNA in the establishment, maintenance and perpetuation of bacterial biofilms. Critical reviews in microbiology. 2015;41(3):341-352.

[78] Karygianni L, Ren Z, Koo H, Thurnheer T. Biofilm matrixome: extracellular components in structured microbial communities. Trends in Microbiology. 2020.

[79] Barnes AM, Ballering KS, Leibman RS, Wells CL, Dunny GM. *Enterococcus faecalis* produces abundant extracellular structures containing DNA in the absence of cell lysis during early biofilm formation. MBio. 2012;3(4).

[80] Edlund A, Yang Y, Yooseph S, He X, Shi W, McLean JS. Uncovering complex microbiome activities via metatranscriptomics during 24 hours of oral biofilm assembly and maturation. Microbiome. 2018;6(1):217.

[81] McLean JS. Advancements toward a systems level understanding of the human oral microbiome. Frontiers in cellular and infection microbiology. 2014;4:98.

[82] Simon-Soro A, Guillen-Navarro M, Mira A. Metatranscriptomics reveals overall active bacterial composition in caries lesions. Journal of oral microbiology. 2014;6:25443.

[83] Peterson SN, Snesrud E, Liu J, Ong AC, Kilian M, Schork NJ, et al. The dental plaque microbiome in health and disease. PloS one. 2013;8(3):e58487.

[84] Duran-Pinedo AE, Chen T, Teles R, Starr JR, Wang X, Krishnan K, et al. Community-wide transcriptome of the oral microbiome in subjects with and without periodontitis. The ISME journal. 2014;8(8):1659-1672.

[85] Bagyi K, Haczku A, Marton I, Szabo J, Gaspar A, Andrasi M, et al. Role of pathogenic oral flora in postoperative

pneumonia following brain surgery. BMC infectious diseases. 2009;9:104.

[86] Rivas Caldas R, Le Gall F, Revert K, Rault G, Virmaux M, Gouriou S, et al. Pseudomonas aeruginosa and Periodontal Pathogens in the Oral Cavity and Lungs of Cystic Fibrosis Patients: a Case-Control Study. Journal of clinical microbiology. 2015;53(6):1898-1907.

[87] Wang Z, Zhou X, Zhang J, Zhang L, Song Y, Hu FB, et al. Periodontal health, oral health behaviours, and chronic obstructive pulmonary disease. Journal of clinical periodontology. 2009;36(9):750-755.

[88] Dickson RP, Erb-Downward JR, Freeman CM, McCloskey L, Beck JM, Huffnagle GB, et al. Spatial Variation in the Healthy Human Lung Microbiome and the Adapted Island Model of Lung Biogeography. Annals of the American Thoracic Society. 2015;12(6):821-830.

[89] Johanson WG, Pierce AK, Sanford JP. Changing pharyngeal bacterial flora of hospitalized patients. Emergence of gram-negative bacilli. The New England journal of medicine. 1969;281(21):1137-1140.

[90] Whiteson KL, Meinardi S, Lim YW, Schmieder R, Maughan H, Quinn R, et al. Breath gas metabolites and bacterial metagenomes from cystic fibrosis airways indicate active pH neutral 2,3-butanedione fermentation. The ISME journal. 2014;8(6):1247-1258.

[91] de Martel C, Georges D, Bray F, Ferlay J, Clifford GM. Global burden of cancer attributable to infections in 2018: a worldwide incidence analysis. The Lancet Global health. 2020;8(2):e180-ee90.

[92] Fitzpatrick SG, Katz J. The association between periodontal disease and cancer: a review of the literature. Journal of dentistry. 2010;38(2):83-95.

[93] Mirvish SS. Role of N-nitroso compounds (NOC) and N-nitrosation in etiology of gastric, esophageal, nasopharyngeal and bladder cancer and contribution to cancer of known exposures to NOC. Cancer letters. 1995;93(1):17-48.

[94] Meurman JH. Oral microbiota and cancer. Journal of oral microbiology. 2010;2.

[95] Chalabi M, Moghim S, Mogharehabed A, Najafi F, Rezaie F. EBV and CMV in chronic periodontitis: a prevalence study. Archives of virology. 2008;153(10):1917-1919.

[96] Kato I, Vasquez AA, Moyerbrailean G, Land S, Sun J, Lin HS, et al. Oral microbiome and history of smoking and colorectal cancer. Journal of epidemiological research. 2016;2(2):92-101.

[97] Fan X, Alekseyenko AV, Wu J, Peters BA, Jacobs EJ, Gapstur SM, et al. Human oral microbiome and prospective risk for pancreatic cancer: a population-based nested case-control study. Gut. 2018;67(1):120-127.

[98] Ali J, Pramod K, Tahir MA, Ansari SH. Autoimmune responses in periodontal diseases. Autoimmunity reviews. 2011;10(7):426-431.

[99] Scannapieco FA, Dasanayake AP, Chhun N. "Does periodontal therapy reduce the risk for systemic diseases?". Dental clinics of North America. 2010;54(1):163-181.

[100] Sharma N, Bhatia S, Sodhi AS, Batra N. Oral microbiome and health. AIMS microbiology. 2018;4(1):42-66. Epub 2018/01/12.

[101] Willis JR, Gabaldon T. The Human Oral Microbiome in Health and Disease: From Sequences to Ecosystems. Microorganisms. 2020;8(2). Epub 2020/02/28.

Chapter 4

Assessing Host-Pathogen Interaction Networks via RNA-Seq Profiling: A Systems Biology Approach

Sudhesh Dev Sareshma and Bhassu Subha

Abstract

RNA sequencing is a valuable tool brought about by advances in next generation sequencing (NGS) technology. Initially used for transcriptome mapping, it has grown to become one of the 'gold standards' for studying molecular changes that occur in niche environments or within and across infections. It employs high-throughput sequencing with many advantages over previous methods. In this chapter, we review the experimental approaches of RNA sequencing from isolating samples all the way to data analysis methods. We focus on a number of NGS platforms that offer RNA sequencing with each having their own strengths and drawbacks. The focus will also be on how RNA sequencing has led to developments in the field of host-pathogen interactions using the dual RNA sequencing technique. Besides dual RNA sequencing, this review also explores the application of other RNA sequencing techniques such as single cell RNA sequencing as well as the potential use of newer techniques like 'spatialomics' and ribosome-profiling in host-pathogen interaction studies. Finally, we examine the common challenges faced when using RNA sequencing and possible ways to overcome these challenges.

Keywords: RNA-Seq, transcriptome, next generation sequencing, systems biology, host-pathogen interactions

1. Introduction

1.1 RNA sequence profiling

RNA sequencing (most commonly abbreviated as RNA-Seq) is an advanced sequencing approach that has transformed the way we look at the intricacies that exist within complex biological systems. Using high-throughput next generation sequencing (NGS) technology, RNA-Seq allows the detection and quantification of RNA transcripts in a biological sample with high accuracy [1]. Further analysis of RNA-Seq data can reveal a dynamic scale of information ranging from alternative spliced transcripts, gene fusions, single nucleotide polymorphisms (SNPs), post-translational modifications, temporal fluctuations in RNA expression during infection across cells [2–5]. This extensive capability of RNA-Seq has also recently found its way into studies investigating host-pathogen interaction networks with hopes of further elucidating this multi-faceted system [6, 7].

One of the earliest papers describing the term 'RNA-Seq' successfully mapped the transcriptome of the yeast genome using a high-throughput sequencing platform [8]. In fact, a handful of studies had already started using the RNA-Seq method even before the term was coined [9–13]. Commonly referred to as 'transcriptome sequencing', these studies mainly adopted the massively parallel pyro-sequencing technology which was one of the newer sequencing technologies at the time [14]. While DNA sequencing and genomic studies have led to many breakthroughs, RNA-Seq brings forth a more functional, integrated view of expressed genes with distinct advantages over previous methods. Different aspects of RNA-Seq will be discussed in the following sections leading to its role in unravelling host-pathogen interaction networks.

2. Introduction to RNA Seq approaches in biology and medicine

Transcriptomics is an area that is being continuously developed especially with the recent advances in technology that make it easier to carry out large-scale analysis of RNA. Prior to the use of RNA-Seq, traditional methods used to study transcriptomes include hybridization-based, sequence-based and tag-based approaches [15]. A popular hybridization-based approach is the use of microarrays. The main principle behind microarrays is complementary binding of nucleotides. A microarray or 'gene chip' is prepared containing thousands of different oligonucleotides or cDNA molecules [16]. Extracted RNA samples converted into cDNA are fluorescently labelled and allowed to hybridise on the microarray [17]. This approach has proven to be useful in studies looking to compare the levels of gene expression but it does not generate quantitative values and can only be used for known genes [18]. A related method called genome tiling array, however, has the ability to examine genomic regions without prior knowledge of its expression [19]. Like any other method scrutinised over time, the pitfall of microarrays stem from inconsistent protocols, high background noise due to cross-hybridization, low technical reproducibility as well as other technical issues [20, 21].

As for sequence-based approaches, a method used for gene discovery early on was expressed sequence tags (ESTs), which are single-pass sequence reads selected from cDNA libraries [22]. Aside from being expensive, the single-pass reads produced using this method are more prone to error and likely to have redundancies in large datasets [23]. On the other hand, tag-based approaches like serial analysis of gene expression (SAGE) and massively parallel signature sequencing (MPSS) employ the principle of generating short 'tags' (9–20 base pairs) which are then sequenced and quantified on a large scale [24, 25]. Both methods make use of bead-based technology and produce accurate quantitative levels of gene expression but mostly focusing on the 3′-ends [26]. Cap analysis of gene expression (CAGE) was then introduced to examine 5′-end short tag sequences revealing more information about promoters and transcription start sites [27]. Altogether, these relatively costly methods were common during the Sanger sequencing era and could only be optimally used in conjunction with already known genome or EST databases. In addition, these approaches had limitations such as cloning biases, technical challenges and general lack of strength to be solid stand-alone approaches for transcriptome analysis [28, 29].

After decades of utilising Sanger sequencing, the development of Next Generation Sequencing (NGS) was a giant leap for researchers everywhere. There has been constant development in NGS technologies hence they can be more distinctly categorised as second-, third- and even fourth generation sequencing.

Second generation sequencing mainly consists of two methods which are sequencing by hybridization (SBH) and sequencing by synthesis (SBS) [30]. SBH was the main principle behind microarray technology using known DNA sequences as explained previously. Meanwhile, SBS is different from Sanger sequencing because dideoxy terminators are not used. In addition, it employs repeated cycles of nucleotide incorporation and also tiny-volume reactions that are massively run in parallel. Most second generation methods commonly rely on sequencing reactions that take place in micro wells or channels [30]. One of the most common second generation sequencing technology is developed by Illumina, producing short read lengths. On the other hand, third- and fourth generation sequencing technologies are more focused on producing longer read lengths. These technologies have creatively exploited the principle of sequencing reactions occurring in millions of tiny wells either by specially engineered chambers or biological nanopores [30]. The front runners of third- and fourth generation sequencing are currently Pacific Biosciences and Oxford Nanopore Technologies. Their technologies will be discussed in the coming sections. Also known as deep sequencing, these high-throughput sequencing technologies eventually led to the development of next generation RNA-Seq. Originally described by Nagalakshmi et al. [8], preliminary RNA-Seq studies focused on improving genomic annotation by examining novel untranslated regions, promoter regions, intergenic transcripts, alternative gene splicing events and single nucleotide polymorphisms (SNPs) among others [31–35]. Advances in next generation RNA-Seq has allowed diverse studies spanning areas like diagnosis of genetic conditions, characterisation of immune microenvironments, understanding cellular frameworks and viral genetics [36–40]. **Table 1** shows a comparison of RNA-Seq with some of the main methods used to study the transcriptome.

	Microarray	SAGE*	Next-Gen* RNA-Seq
Type of method	Hybrid-based	Tag-based	cDNA library preparation & high throughput sequencing
Amount of input material	High	High	Low
Probes	Yes	No	No
Cost	Medium	Low	High
Data analysis	Based on relative intensity	Based on amplified SAGE tag counts	Based on amplified & sequenced cDNA fragments producing raw read counts
Detection of novel genes/transcripts	No	Limited	Yes
Detection of alternatively spliced isoforms	Limited	No	Yes
Detection of single nucleotide polymorphisms	No	No	Yes
Detection of non-coding transcripts	Limited	Limited	Yes
Prior knowledge of gene sequence	Yes	Limited	No

**SAGE – Serial Analysis of Gene Expression.*
**Next-Gen – Next Generation.*

Table 1.
Comparison of commonly used methods for gene and transcriptome analysis.

2.1 Experimental flow in approaches

The flow chart in **Figure 1** shows the initial steps involved when carrying out an RNA-Seq experiment.

The first step in an RNA-Seq experiment is to isolate RNA from any biological sample (e.g. cell or tissue populations). As a quality control step, the integrity of extracted RNA samples is commonly measured using an Agilent Bioanalyzer. Based on electrophoretic separation of RNA and a built-in software algorithm, it produces an RNA Integrity Number (RIN) depicting levels of RNA degradation [43]. The next step involves either an enriching or depleting procedure to select specific RNA species. In any given total RNA sample, a variety of RNA species would be present including messenger RNAs, ribosomal RNAs, precursor RNAs, non-coding RNAs,

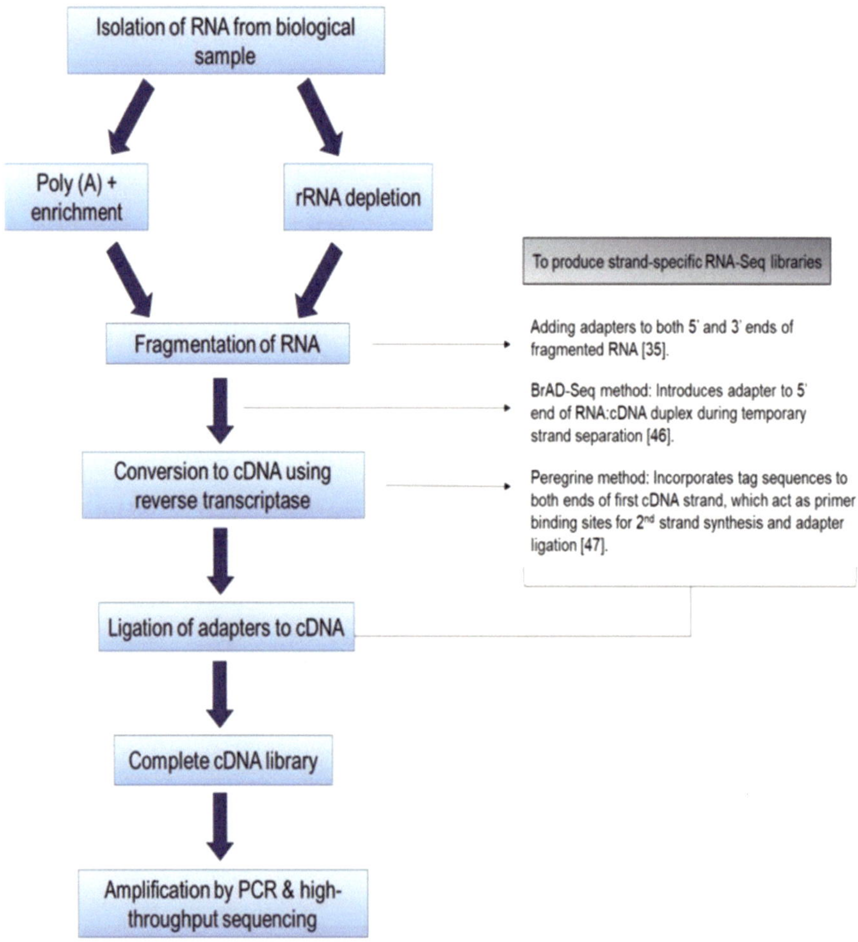

Figure 1.
Overview of second generation RNA-Seq workflow. Firstly, RNA samples are extracted from biological samples. Selection of specific RNA species is carried out either by enriching transcripts expressing poly-adenylated (poly-A) tails (usually mRNA) or by removing the abundant ribosomal RNAs (rRNAs). Next, the enriched or depleted RNA samples are fragmented followed by reverse transcription to generate cDNA. The next step is ligation of adapters, however, standard adapter ligation loses information about RNA strand-specificity hence a few methods have been developed to prevent this. These include adding adapters directly to the 5′ and 3′ ends of fragmented RNA [31], the BrAD-Seq method which adds an adapter to 5′ end of the RNA:cDNA duplex during reverse transcription [41], and lastly the Peregrine method which incorporates tag sequences to 5′ and 3′ ends of the first cDNA strand [42]. Once library preparation is completed, samples are amplified by PCR and RNA-Seq library is now ready to be sequenced.

etc. A bulk of the RNA portion (~95%) in most cells comprises of rRNA which if not removed, would make up a large part of the sequencing reads. Since this would largely restrict the study of less-abundant RNA species, protocols were created to circumvent this issue.

One such protocol is the enrichment of polyadenylated (poly-A) RNAs. This procedure selects for poly (A) + RNA mainly mRNA and exploits the fact that rRNAs generally lack this structure. A particular study however did find the presence of rRNA polyadenylation but only in small amounts [44]. This selection step can be carried out by using magnetic beads coated with oligo-dT or reverse transcription (RT) using oligo-dT primers [45]. An alternative step is rRNA depletion which serves to eliminate them from total RNA samples. There are various approaches used by different researchers for this method. One such approach uses probes like biotinylated DNA or locked nucleic acid which are allowed to hybridise to rRNAs. This is followed by a depleting step using streptavidin beads [46]. Another method that can be used for rRNA depletion is known as probe-directed degradation (PDD). This method involves obtaining cDNA:RNA duplexes, circularising them and then hybridising them with rRNA-specific probes. The final step involves digestion with Duplex-Specific Nuclease (DSN) which renders the hybridised-sequences unusable [47]. Some researchers also use not-so-random (NSR) primers that bind to specific RNA molecules during RT, excluding rRNAs [48]. In essence, the variety of methods that exist for rRNA depletion focuses on unique features of rRNA that can be singled out and developed into an eliminating step. The choice of using either poly (A) + selection or rRNA depletion ultimately depends on the aims of the experiment. Evaluation of these two methods showed that while rRNA depletion could record more unique characteristics of the transcriptome, poly(A) + selection was more accurate in terms of gene quantification [49].

Following poly (A) + enrichment or rRNA depletion, RNA samples need to be fragmented to shorter sequences according to the size restrictions of sequencing platforms. RNAs are usually fragmented chemically using alkaline solutions, divalent cations or enzymes [45]. Alternatively, RNA can be reverse transcribed (RT) first followed by cDNA fragmentation. Similarly, enzymes like DNAses can be used to fragment cDNA with recent advances including a transposon-based approach [50]. Next, either fragmented RNAs or cDNAs are ligated with adapters that are specific to the sequencing platform to be used. This step however overlooks RNA directionality whereby there is lack of information about DNA strands and their corresponding sense RNA strands. This may impede the identification of novel RNA species and also make it harder to accurately measure sense RNA expression [45]. Methods have been developed to preserve this directionality and they can be carried out either directly on fragmented RNA, cDNA or even on RNA:cDNA hybrids that are formed during RT. One of these approaches include adding distinct adapters to the 5′ and 3′ ends of fragmented RNA [31]. This difference in sequences at both ends preserve the strandedness of RNA. Other methods to preserve strand-specificity of RNA are BrAD-Seq [41] and the Peregrine method [42]. The BrAD-Seq method exploits the transient strand separation or 'breathing' of RNA:cDNA hybrid during reverse transcription to add an adapter to the 5′ end of the duplex. This is followed by incorporation of nucleotides by *E.coli* DNA Polymerase I to form the second strand and eventually a complete strand-specific cDNA library. Meanwhile, the Peregrine method incorporates short unidentical tag sequences to the ends of cDNA during first strand synthesis. These then serve as primer binding sites for subsequent adaptor ligation during second strand synthesis.

Finally, after cDNA synthesis and adapter ligation, cDNA libraries need to be amplified using PCR. Once amplified, they are ready for sequencing using a chosen NGS sequencing technology.

3. Next-generation sequencing technologies

3.1 Illumina, second generation sequencing technology

In 2005, Solexa released the Genome Analyser which established a quality standard for the transformation of sequencing platforms that came after. Solexa was bought over by Illumina in 2007 and continued developing second-generation sequencing platforms for specific aims [51]. The strategy behind Illumina's sequencing process is a four-colour reversible termination sequencing method. After clonal amplification of DNA, sequencing occurs through base incorporation onto the template strand successively, followed by washing, imaging and cleavage. In this method, the polymerisation reaction is halted using fluorescently-labelled dNTPs and unincorporated bases are removed. Final analysis is carried out on the obtained four-colour images to ascertain base composition [52]. Currently, Illumina provides an impressive number of sequencing platforms which include MiniSeq, MiSeq, NextSeq 550, NovaSeq 6000, etc. NextSeq 500 was discontinued with the introduction of NextSeq 550 which has more flexible features of microarray scanning and sequencing. Their newest sequencing systems, NextSeq 1000 and 2000, boasts an integrated cartridge containing fluidics, waste compartment and reagents. It also possesses a novel system taking advantage of super resolution optics resulting in higher sensitivity and increased accuracy of imaging data [53].

3.2 Pacific Biosciences, third generation sequencing technology

The single-molecule real-time sequencing (SMRT) method is a third-generation sequencing approach developed by Pacific Biosciences (PacBio). This method directly observes DNA or cDNA synthesis by DNA polymerase as it occurs in real time [54]. The principle behind this method is the use of zero-mode waveguide (ZMW) technology. A ZMW is essentially a tiny, zeptoliter-sized hole deposited slightly above a glass surface [54]. Within each ZMW is a chamber containing a single DNA polymerase molecule affixed to the bottom glass surface using a biotin/streptavidin system. Fluorophore-labelled nucleotides are added to the compartment above an array of ZMWs. Diffusion then occurs whereby labelled nucleotides travel downwards through the ZMW to reach DNA polymerase for incorporation onto the DNA strand. The ZMW system is sufficiently sensitive to detect incorporations against background nucleotides. In addition, one of the first commercially available sequencing system employing SMRT contains an assembly of ~75000 ZMWs [54]. Therefore, single-molecule sequencing can be carried out massively in parallel. As of now, PacBio also has an Iso-Seq method used to analyse long reads produced by SMRT to examine novel transcripts, gene fusion, alternative splicing, etc. Their newest system release is the sequel IIe system that promotes higher quality data, shorter analysis time and cheaper costs [55].

3.3 Oxford Nanopore Technologies, fourth generation sequencing technology

As suggested by their name, Oxford Nanopore Technologies (ONT) developed and commercialised nanopore-based sequencing. The idea behind this strategy is that each nucleotide can induce a unique fluctuation in ionic current while passing through a tiny channel [56]. An α-hemolysin pore secreted by *Staphylococcus aureus* was used to form single transmembrane channels through which nucleic acid polymers would pass through [56]. This study aimed to determine the length of nucleic acid polymers but also proposed that if each nucleotide could provide a characteristic current change based on their chemical or molecular properties, it could very well be used to determine nucleotide sequences as well. The current technology employed by ONT consists of a group of tiny wells contained in a sequencing flow cell. Within each well is a synthetic bilayer fabricated with biologic nanopores. As described earlier, sequencing is achieved by assessing the distinct current changes induced during base incorporation carried out by a molecular motor protein [57]. Presently, the devices provided by ONT include the Flongle, MinION, GridION and PromethION. Flongle and MinION are more for smaller scale experiments while GridION generates high-throughput data up to 150GB. PromethION, on the other hand, provides ultra- high-throughput data of up to a remarkable scale of 8 TB [58].

3.4 Other genome analysers

Roche 454 pyrosequencing was the first commercially successful 2nd generation sequencing platform, initially developed by 454 Life Sciences and later acquired by Roche. Sequencing by this platform depended on the detection of visible light produced by a group of enzymes correlating to the pyrophosphate release during nucleotide incorporation [59]. Roche however stopped supplying the 454 sequencing machines and any accompanying reagents since 2016 [51]. Another NGS instrument is Sequencing by Oligonucleotide Ligation and Detection (SOLiD) released by Applied Biosystems Instruments (ABI). This technology uses sequencing by ligation. It involves cycles of annealing and ligation of primers and probes. Four-colour imaging is also carried out after which ligated probes are cleaved to allow another cycle of ligation [60]. Despite being quite accurate, it has a long run time and requires experts to analyse raw data [51]. Furthermore, another sequencing approach called DNA nanoball sequencing was developed by Complete Genomics and later acquired by Beijing Genomics Institute (BGI) [51]. This approach combines the principles of hybridization and ligation. DNA nanoballs are produced by amplifying DNA or cDNA using rolling-circle replication. They are then added onto a flow cell with an array of wells and each nanoball in each well are sequenced at high density. This process only yields short reads however and takes a long time. Meanwhile, Ion Torrent technology introduced by the team behind the 454 sequencer is based on the electronic detection of pH changes as opposed to detection of light as previously used [61]. Each incorporated nucleotide generates an electronic signal detected by electronic sensors placed at the bottom of each flow cell [51]. Lastly, a third generation sequencing platform called Helicos sequencing employs the principle of single-molecule fluorescent sequencing [62]. The Helicos sequencer, Heliscope, does not require clonal amplification and uses a very sensitive fluorescence detection system [60]. This method merges sequencing by synthesis and hybridization.

3.5 NGS advantages

All NGS platforms have significant advantages over previously used methods however, each platform has their own strengths and unique features. The four major sequencing platforms being used currently are Illumina, Pacific Biosciences, Oxford Nanopore Technologies (ONT) and Ion Torrent. Both Illumina and Ion Torrent are highly accurate but they are relatively more costly and have short reads (≤ 400). The problem with short read lengths is that it prevents researchers from performing de novo assembly and impedes the detection of structural variations [63]. On the other hand, PacBio and ONT platforms produce long reads (≥ 500) but they have variable accuracies. Although, both ONT and PacBio have similar read lengths, ONT specifically the MinION device, has higher error rates of up to 38.2% [64]. ONT also produces a higher yield but PacBio has better data quality overall [65].

All these platforms have a similar disadvantage which is a long turnaround time except for ONT. In addition, ONT also has lower capital costs compared to the others [66].

Illumina sequencing has error rates of <1% and one of their systems called the NextSeq 550, employs the use of two-channel sequencing strategies instead of the four-channel strategy used by previous systems. This method only needs two images to detect nucleotides which makes data processing much faster [67]. However, a few studies found that PacBio sequence data produced better results than Illumina datasets specifically when used for *de novo* assembly purposes in addition to improved resolution [68–70]. Meanwhile, when comparing Illumina against ONT, ONT proved to have a significantly shorter turn-around time of <15 hours while Illumina analysis took around 3 to 6 days. Therefore, ONT sequencing was deemed more suitable for urgent, smaller scale sequencing requirements especially during public health emergencies [71]. Lastly, the Ion Torrent Personal Genome Machine (PGM) has a unique plus point which is the ability to identify single nucleotide polymorphisms (SNPs) better than Illumina and PacBio [72]. Lahens et al. [73] did however conclude from his experiments that both Illumina and Ion Torrent are equally capable in detecting differential gene expressions. There are a large number of studies that have found certain platforms to perform better than others, however it ultimately depends on the aims of the experiment. Another useful method is combining datasets from more than one platform to acquire a more complete genome assembly [74–79].

NGS technologies are also capable of producing either single-end or paired-end reads during sequencing. The question that normally arises is which type of sequencing to perform. Single-end sequencing in RNA-Seq is when a cDNA fragment is sequenced from only one end whereas paired-end sequencing is when both ends of a fragment are sequenced [80]. Paired-end sequencing produces twice the amount of data which increases the accuracy of read alignment. It also more sensitive and allows the detection of events like gene fusions and new splice isoforms. On the other hand, single-end sequencing is much cheaper than paired-end sequencing. It is also more suitable for some methods such as ChIP-Seq and small RNA-Seq [80]. Although it is the more economical choice, it has drawbacks such as lower read counts per RNA feature and a weaker ability to assign reads to features. In the context of functional profiling, single- and paired-end reads in an RNA-Seq experiment only showed a 65% agreement in the top 20 gene ontology (GO) terms obtained. However, when looking at the top 300 GO terms, both led to similar broad conclusions [81]. Since the cost of sequencing is an important consideration to make, Corley et al. [81] suggested that single-end sequencing could be carried out with more biological replicates as they found that it was comparable to the results obtained using paired-end sequencing if functional analysis is done cautiously. As mentioned before, the utility of single- or paired-end sequencing ultimately comes down to the research question. For instance, if the main objective of the experiment is transcriptome assembly, then paired-end sequencing would be the more suitable choice.

4. Application of systems biology in understanding host-pathogen interactions

Systems biology is the comprehensive study of a biological system encompassing molecular- level interactions, sub-cellular dynamics and overall physiological functions of cells, tissues and organs [82]. A systems biology approach aims to looks at the larger picture involved in a given system or condition. For a long time, research had centred on the molecular understanding of genes and proteins. Current illustrations or diagrams of interconnecting pathways are just not enough to completely understand a system. Kitano et al. [83] aptly describes these diagrams as mere static roadmaps, whereas what we seek to understand leans more toward patterns, their causes and regulatory dynamics. In the context of host-pathogen interactions, a systems biology view is examining components from both the host and pathogen as well as their interactions with one another. Some of the approaches used in systems biology include identification of key molecules or biomarkers, inference between networks and disease module discovery [84]. The advancement of -omics technologies supported by high throughput sequencing has increased the whole-system analyses focusing on host-pathogen interaction between genes, proteins and small ligands [85]. This is accomplished by carrying out dual RNA sequencing whereby both host and pathogen transcriptomes are profiled during the course of an infection. Multiple cascades of events are triggered by an infection and dual RNA-Seq allows the monitoring of host and pathogen in parallel. Knowledge gained from comprehensive host-pathogen interaction studies especially with the use of dual RNA-Seq can guide efforts toward better therapeutics against infection. Dual RNA-Seq was first described by Westermann et al. [86] however it only started gaining attention recently resulting in a surge of studies utilising this method.

4.1 Bacteria-host interaction

Interaction between bacteria and hosts usually begin with a compulsory attachment or adherence of bacteria to host cells followed by subsequent internalisation which may involve direct or indirect receptor binding [87]. Entry into the host may seem like a straightforward step but it involves a drastic change in environment for the pathogen. Hence, entry and any subsequent mechanism employed are bound to involve a complex interplay between the host and pathogen. Previous methods were limited in the sense that they only allow the analysis of mRNA in either infected host cells or bacteria [88]. Dual RNA-Seq has provided researchers everywhere an access to the complete story. Some of the host-bacteria interaction studies utilising dual RNA-Seq have looked at bacteria infecting humans, such as *Salmonella enterica* [89], *Haemophilus influenza* [90], *Streptococcus pneumonieae* [91, 92], *Mycobacterium tuberculosis* [93, 94] and *Mycobacterium leprae* [95]. Despite the diversity of these bacteria-host dual RNA-Seq studies, one similarity is that all their findings encompass several aspects or levels of a biological system instead of mere isolated observations. For instance, in the study by Baddal et al. [90], not only did they characterise preferential binding of nontypeable *H.influenzae* (NTHi) to ciliated bronchial epithelial cells, they also observed differential expression of various bacterial virulence factors, alteration of host cell adherence junctions, host-dependent modulation of NTHi metabolic machinery and rearrangement of host extracellular matrix and cytoskeletons. In addition, they discovered small RNA regulatory elements that were differentially expressed including novel snoRNAs that have never been associated with NTHi before. Meanwhile, Aprianto et al. [91] observed the generation of reactive oxygen species (ROS) by *S. pneumoniae*, the glutathione-dependent detoxification of ROS as a counteraction by the host, expression of chemokine IL-8 for immune response repression and also the activation of bacterial sugar transporters sensitive to host-derived non-glucose carbohydrates. Lastly, Yimthin et al. [96] analysed the whole blood transcriptome of 29 patients with melioidosis which is the infection caused by *B. pseudomallei* often leading to mortality in endemic areas. Using RNA-Seq, they managed to identify survivor- and non-survivor-specific expressions related to cell lineage processes and immune activation pathways with the potential to be biomarkers against melioidosis. These findings further reiterate the importance of a systems biology-based view when analysing RNA-Seq data spanning multiple gene networks and pathways.

4.2 Virus-host interaction

Viruses are obligate intracellular parasites manipulating various machinery and components of the host cell. The human body has developed efficient responses against viruses particularly the interferon system. An antiviral state is induced by the family of interferon proteins and other effectors upon viral infection. However, over time, certain viruses have evolved mechanisms to dodge these immune responses [97]. Given the complex nature of viral infections, it most certainly involves multi-level interactions and a method like dual RNA-Seq can help us understand these elaborate interactions networks. One of the first studies examining host-virus interactions using dual RNA-Seq was carried out using a murine infection model for cytomegalovirus (CMV) [98]. This study found some unexpected results such as highly abundant viral transcripts with unknown functions and also a viral transcript bearing functions of both non-coding RNA and mRNA. From the host perspective, expected upregulation of genes involved in inflammation and immunity were observed. Certain unforeseen results include upregulation of genes associated with development and differentiation. More importantly, this study found many differentially expressed genes within specific biological pathways including certain networks with unknown relevance to infection, providing new insights into CMV pathogenesis. The use of dual RNA-Seq has been applied to a range of studies analysing host-virus interactions which include infections by avian influenza (H5N8) [99], varicella zoster virus [100], Crimean-Congo hemorrhagic fever virus (CCHFV) [101], influenza A (H3N2) [102], and Zika virus [103]. Similar to host-bacterial studies, a wide range of findings were uncovered including variable alternative gene splicing events, association between clinical phenotypes and viral gene induction, remodelling of host epidermal environment, inhibition of functional pathways, host metabolic regulation and many more. Michlmayr et al. [103] successfully identified CD169 (Siglec-1) on CD14+ monocytes as a potential biomarker against acute infections of Zika virus while also providing evidence that dengue-immune patients did not necessarily have an upper hand when faced with Zika virus. Another interesting study by Wesolowska-Andersen et al. [104] using dual RNA-Seq found that transcriptionally active respiratory viruses were present in children even in the absence of any observable respiratory illness. These viral carriers also displayed alterations in their nasal transcriptomes. This shows that underlying host-virus interaction networks are still being engaged 'silently' and not necessarily in cases where the illness clearly manifests itself. In due time, these studies will hopefully reveal horizontal inter-study patterns which will point toward the discovery of common disease modules or host-pathogen interaction networks. Furthermore, the discovery of a novel coronavirus in Hong Kong was achieved through a series of eliminating laboratory tests and eventually genome sequencing [105]. In addition to discovery of novel pathogens, RNA-Seq analysis can provide information relating to genome sequence, gene expression, pathogen abundance and a myriad of information that will provide useful insight regarding the pathogen and how it causes disease [106]. Currently, most RNA-Seq studies examining novel viruses are focused on plant viruses [107, 108]. The rapid detection of novel viruses in humans by RNA-Seq is an area that should be further investigated and optimised as it can help us take precautionary steps before the wide spread of disease.

4.3 Fungi-host interaction

There are at least 712 000 existing fungal species around the world however the total number of fungal species is estimated to be more than 1.5 million [109]. The proportion of fungal species causing human diseases are quite small comparatively [110]. Some of the most common opportunistic fungal pathogens are *Aspergillus fumigatus* and *Candida albicans*. Previous studies have elucidated certain interactions of these fungi with their host including interference of host phagolysosome mechanisms, activation of complement system, morphological switches and formation of neutrophil extracellular traps (NETs) [111–113]. These studies mainly use assay- and imaging-based techniques to study interaction and are mostly focused on specific pathways or components. From a systems biology perspective, pathogenic fungi often co-evolve with the host and commensals resulting in an equilibrium shift within the host leading to a myriad of changes affecting many networks [114]. The use of RNA-Seq has allowed a more comprehensive study of host-fungal interactions. Initially, a number of studies used RNA-Seq to delineate transcriptional landscapes for fungi like *Candida albicans* and *Candida glabrata* [115, 116]. In terms of host-fungal interaction, RNA-Seq has shed light on alternative splicing events during host invasion, gene expression profiles in mice models of fungal keratitis and also differences in regulatory networks between *Candia albicans* and *Mus musculus* [117–119]. Dual RNA-Seq analysis of *Trichophyton rubrum*-infected human keratinocytes also demonstrated the upregulation of genes increasing the efficiency of nutrient uptake, production of keratinolytic proteases as well as host-derived antimicrobial proteins [120].

4.4 Combination of pathogens and host interactions

Aside from the pathogens discussed above, some other pathogens that exist are parasites, prions and in rare cases, algae [121–123]. Parasites in particular have extremely complex life cycles involving different hosts at different life stages [124]. A clear comprehension of parasitic life cycles will undoubtedly require a systems biology approach and RNA-Seq has provided an avenue for that. RNA-Seq studies have allowed inter-sex, inter-stage and inter-host studies involving parasites like *Plasmodium falciparum* [125], *Trypanosoma vivax* [126], *Brugia malayi* [127], *Trichuris trichiura* [128] and *Schistosoma mansoni* [129]. A dual RNA-Seq study examining the interactions between murine hosts and the parasite *Toxoplasma gondii* also provided many insights into acute and chronic infection stages by this parasite that is prevalent in humans [130]. Prions, which are misfolded proteins, cause several neurodegenerative diseases in humans including Jakob-Creutzfeldt disease, kuru and fatal familial insomnia [122]. Despite being a protein-only infection, it involves extensive processes occurring simultaneously in the brain including synaptic alterations, inflammation, neural cell death and protein aggregation [131]. RNA-Seq has revealed unique miRNA profiles produced by components of prion-infected cells, mechanisms of prion-induced neurotoxicity and signature glial gene expressions among others [132–134]. Meanwhile, algal infections in humans are quite rare however they have been documented such as human protothecosis caused by the *Prototheca* species [121]. Genome sequencing studies have been carried out to study the sequence and expression of these species, however the use of RNA-Seq in this area is still scarce [135, 136]. There are also cases of co-infections whereby more than one pathogen infects a host simultaneously. Transcriptomic profiling studies of co-infections have shed some light on disease mechanism, molecular phenotypes and inter-disease relationships. One example of a complex co-infection is when HIV-infected patients develop cryptococcal meningitis which is a fungal infection.

Some patients undergoing treatment for these infections also start to develop paradoxical cryptococcosis-associated immune reconstitution inflammatory syndrome (C-IRIS) characterised by various clinical deteriorations. By assessing the whole blood transcriptome of infected patients, Vlasova-St. Louis et al. [137] identified novel and unique biomarkers for both early and late stages of C-IRIS which are difficult to distinguish due to their similar clinical manifestations. Moreover, an ambitious study by Seelbinder et al. [138] managed to carry out a triple RNA-Seq analysis in host monocyte-derived dendritic cells infected by the fungus, *Aspergillus fumigatus* and human cytomegalovirus (CMV). These two pathogens are commonly co- occurring pulmonary pathogens. A highlight from their comprehensive study is that host expression levels that were upregulated during single infection by either pathogen were downregulated instead during co-infection. This implied interference or opposing effects of the two distinct host responses induced and also a possible synergistic relationship between *A. fumigatus* and CMV.

5. Bioinformatics and statistical approaches in analysing RNA-Seq data

The initial experimental workflow of RNA-Seq has been described earlier which briefly include depletion of rRNA or enrichment of mRNA, fragmentation of samples and subsequent reverse transcription to form a cDNA library. These cDNA fragments are then sequenced using a high-throughput sequencing platform. This section will describe the data analysis of RNA-Seq data including statistical approaches taken to analyses differentially expressed genes. The whole process is simplified in **Figure 2**, covering all the important analytical steps involved.

Once sequencing data is obtained in the form of raw reads, quality control and sequence filtering need to be carried. This is a key pre-processing step because next-generation sequencing data may contain unexpected artefacts, poor quality reads, low-complexity regions, high GC content and sequencing errors [139, 140]. The presence of these low-quality sequences will further effect downstream analysis leading to inaccuracies in overall RNA-Seq data interpretation. There are a variety of tools that can be used to perform data pre-processing. Two important pre-processing concepts are the quality assessment of reads and also processing/filtering to remove contaminants, adapter sequences, low-quality sequences [141]. Some of the methods developed include FastQC [142], RSeQC [143], NGSQC [144], Trimmomatic [145] and CutAdapt [146]. Weaknesses of these tools include the inability to carry out both data quality control and processing steps, slow run times and single-platform services [147, 148]. Recently developed tools are more comprehensive, encompassing all steps required in raw reads processing. Some of these include FastProNGS [147], FastqPuri [149], Zseq [140], RNA-QC-Chain [150] and fastp [151].

The next step is mapping or aligning the quality-assessed reads onto a genome or transcriptome. Reads can be mapped either uniquely to a single position or multiple positions (multi-reads) in the reference genome. Some of the mapping software or algorithms available are STAR [152], TopHat2 [153], MapSplice [154], BowTie2 [155] and Magic-BLAST [156] among others. A range of bench-marking studies have compared the efficiencies of various RNA-Seq aligners. Baruzzo et al. [157] examined 14 common RNA-Seq aligners, whereas Schaarschmidt et al. [158] evaluated 7 alignment tools. In addition, Engstrom et al. [159] carried out comprehensive analysis on a total of 26 alignment protocols. A similarity across these three studies is that they all found STAR to be one of the more reliable aligners, although other aligners do have their own strengths. After alignment, transcript identification

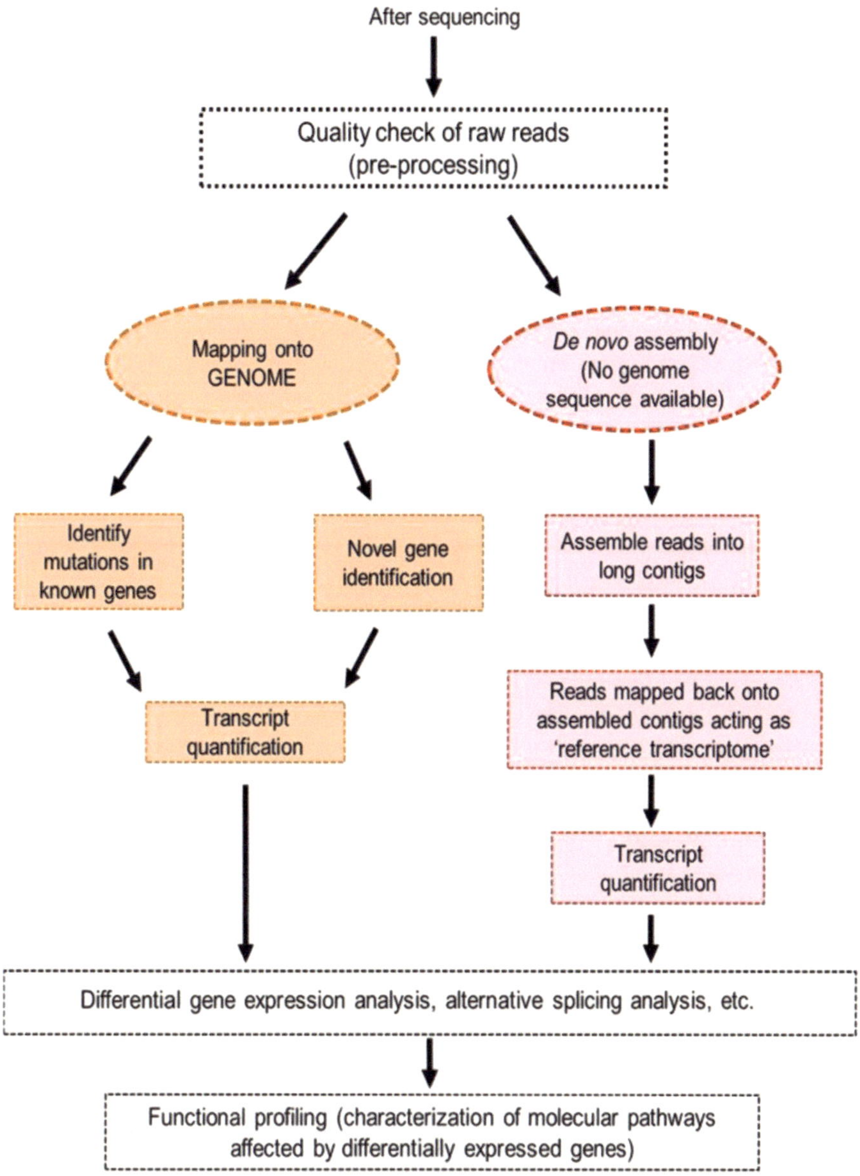

Figure 2.
General RNA-Seq data analysis workflow. The first step after sequencing is pre-processing the sequence reads to obtain data with higher quality. Reads can be either mapped onto a reference genome (e.g., GRCh38) or in cases where a reference genome is unavailable, de novo assembly is carried out. When using a reference genome, novel transcript discovery is possible. After identification of relevant transcripts, quantification or counting is carried out. When the genome sequence is unavailable, de novo assembly is used to assemble reads into long contigs. Reads are then mapped back onto assembled transcriptome followed by quantification. In both cases, differential gene expression and alternative splicing analysis can be carried out in addition to other methods depending on the experiment. Finally, functional profiling is done to characterise molecular pathways and interactions.

is carried out. Reads that are mapped onto known reference transcriptomes can only focus on quantification and not novel transcript discovery. Meanwhile, reads mapped onto a reference genome can either be identified as known transcripts or alternative transcripts [139]. For rapid discovery of novel transcripts, a popular programme called Cufflinks utilises existing annotated genomes as a reference to assist in transcript assembly [160]. Other methods focusing on novel transcript identification are SLIDE [161], iReckon [162] and StringTie [163]. In the case where a reference genome is absent or incomplete, de novo transcript reconstruction is carried out. Reads are first assembled into longer contigs, then this is treated as the 'reference transcriptome' to which the reads are mapped back onto for quantification purposes. Some of the tools available for de novo transcript assembly include Trinity [164], SOAPdenovo-Trans [165], TransABySS [166] and Oases [167]. Depending on the experiment, transcript identification and quantification can be carried out either simultaneously or sequentially. One of the most frequent applications of RNA-Seq is estimating the abundance of gene or transcript expressions. HTSeq-count and featureCounts are two gene-level quantification approaches with HTSeq-count being specially designed for downstream differential expression analysis [168, 169]. These are 'union exon'-based approaches whereby exons that overlap are merged to form a union-exon. This method can assign reads to respective genes with high confidence however, difficulty arises when dealing with alternatively spliced transcripts [170]. Due to biases related to transcript length and number of reads, within-sample normalisation methods are used to standardise reads with some common measures like RPKM (reads per kilobase of exon model per million reads), FPKM (fragments per kilobase of exon model per million mapped reads) and TPK (transcripts per million) [34, 139]. Besides union exon-based methods, several transcript-level statistical quantification methods also exist such as RSEM [171], eXpress [172] and TIGAR2 [173]. Recently, alignment-free methods have also been developed like Salmon [174], kallisto [175] and Sailfish [176].

A crucial step before carrying out differential gene expression (DGE) analysis is data normalisation. The within-sample normalisation approaches during quantification are not sufficient in cases where high numbers of differentially expressed transcripts exist [139]. The current software that exist for RNA-Seq differential gene expression analysis can be mainly categorised into four groups based on the statistical methods employed [177]. These include (1) Poisson or negative binomial model-based methods – baySeq [178], DESeq [179], DESeq2 [180], EBSeq [181], edgeR [182], NBPSeq [183], PoissonSeq [184], TSPM [185], (2) *t*-test analogical methods – Cuffdiff [186], Cuffdiff2 [187], (3) non-parametric methods – NOIseq [188] and SAMseq [189], (4) linear models – limma [190] and voom [191]. Other methods have also been developed including a hybrid full Bayes-empirical Bayes method (ShrinkSeq) and also and binomial distribution-based method called DEGSeq [192, 193]. There are also specific methods that have been developed to study differential gene expression using de novo transcriptome assemblies [194]. There is still no consensus as to which methods are significantly superior however many studies have done comparative analyses of these methods. **Table 2** summarises past studies that have compared the ability of various statistical methods.

A common finding across these studies is that no single method is superior in all circumstances. Each method has their own strengths and weaknesses. Out of the seven studies mentioned in **Table 2**, edgeR and DESeq were commonly found to perform better than other softwares however, a few studies did find contrasting results. Ultimately, the choice of statistical approach largely depends on the nature of study, type of biological sample, number of replicates, budget of study and many other factors that need to be matched to the strengths of any particular approach.

The next step usually examines differential gene expression at a transcript level which is alternative splicing (AS) events. Many computational tools exist that can infer AS events including some of the previously mentioned methods [202]. These include exon-based methods like DEXSeq [203] and JunctionSeq [204], event-based methods like MAJIQ [205], dSpliceType [206] and SUPPA2 [207] and lastly isoform-based methods like Cuffidiff2 [187] and DiffSplice [208]. The final step is a pathway enrichment analysis. The list of DEGs obtained are further analysed

Author (Year)	Statistical methods compared	Data used	Main Findings
Robles et al. [195]	DESeq, edgeR, NBPSeq	Simulations using statistical models derived from real RNA-Seq data	• DESeq performs more conservatively • More biological replicates result in higher quality and reliability of DEG detection
Soneson & Delorenzi [196]	baySeq, DESeq, EBSeq, edgeR, NBPSeq, NOIseq, SAMseq, ShrinkSeq, TSPM, voom+limma, vst + limma	Simulations using statistical models derived from real RNA-Seq data	• voom+limma and vst-limma performed well under many conditions like detection of DEGs, gene ranking and detection of true positives. • SAMseq did well with large sample sizes • TSPM most affected by sample size
Rapaport et al. [197]	baySeq, Cuffdiff, DESeq, edgeR, limma, PoissonSeq	Used benchmark datasets: SEQC dataset & ENCODE project data	• Negative binomial methods (baySeq, DESeq & edgeR) have better specificity, sensitivity & good control of false positive errors • Cuffdiff had low specificity, sensitivity & high false positives • Number of sample replicates greatly affect DEG detection accuracy.
Zhang et al. [198]	Cuffdiff2, DESeq, edgeR	Real RNA-Seq & simulated datasets: MAQC dataset (human), K_N dataset (mouse), LCL dataset (human)	• edgeR performs better than Cuffdiff2 & DESeq in uncovering true positives • Cuffdiff2 more sensitive to sequencing depth, DESeq more sensitive to unbalanced sequencing depths between groups • All three perform better with biological/technical replicates
Seyednasrollah et al. [199]	baySeq, Cuffdiff2, DESeq, EBSeq, edgeR, limma, NOIseq, SAMseq	Real mouse RNA-Seq and human RNA-Seq data	• DESeq & limma most reliable choices • edgeR had large variability, SAMseq had low power • Cuffdiff2 & NOIseq did not do well with large replicates
Rajkumar et al. [200]	Cuffdiff2, DESeq2, edgeR, TSPM	Real RNA-Seq data from mice amygdalae micro-punches	• edgeR had relatively high sensitivity & specificity • Cuffdiff2 had high false positive rates • DESeq2 & TSPM had high false negative rates • RNA sample pooling is discouraged due to low positive predictive values
Costa-Silva et al. [201]	baySeq, DESeq, DESeq2, EBSeq, edgeR, limma+voom, NOIseq, SAMseq	Real RNA-Seq dataset produced for MAQC project	• DESeq2, limma+voom & NOIseq produced most consistent results in terms of accuracy, precision & sensitivity

Abbreviations: TSPM: Two-stage Poisson Model, DEG: Differentially expressed genes, SEQC: Sequencing Quality Control, ENCODE: Encyclopaedia of DNA Elements, MAQC: MicroArray Quality Control, LCL: Lymphoblastoid cell line.

Table 2.
A compilation of numerous studies that have compared common statistical methods used for differential gene expression analysis in RNA-Seq.

to characterise their molecular involvement in biological pathways. Some of the RNA-Seq-specific tools developed for this aim are GOSeq [209], Gene Set Variation Analysis (GSVA) [210] and SeqGSEA [211]. Annotation databases such as KEGG [212], Gene Ontology [213] and Bioconductor [214] also complement functional profiling of DEGs. This is an important step particularly in host-pathogen interactions to unravel the interaction networks that exist. Common databases and softwares used by dual RNA-Seq studies examining host-pathogen interactions are Gene Ontology and KOBAS (KEGG Orthology-based Annotation System) [215–217]. Novel transcripts detected based on de novo assembly can be functionally annotated by finding orthologous proteins in protein databases. Challenges arise when annotating non-protein coding transcripts like long non-coding RNAs which still lack proper functional-annotation procedures [139].

6. Other applications of RNA-Seq in host-pathogen interaction studies

RNA-Seq can be applied in very innovative ways to answer many of the questions and mysteries posed by biology and disease. Initially, it was used for simpler research goals like profiling transcriptomes and monitoring gene expression. Over time, RNA-Seq technology has developed rapidly and one of its vital uses is characterising host-pathogen interaction networks. Dual RNA-Seq in particular has been applied to many infection models ranging from bacteria, virus, fungi and parasites as described in previous sections. Understanding the mechanics of infection induced by pathogens and subsequent host response is a crucial step required before proceeding to figure out clinical treatment strategies. Besides utilising dual RNA-Seq, as extensively detailed earlier, another application of RNA-Seq is single cell RNA sequencing (scRNA-Seq). The difference between bulk RNA-Seq and scRNA-Seq is that the latter allows transcriptional comparison of single-cell populations and has the ability to capture cellular heterogeneity that is normally obscured by bulk RNA-Seq [218]. In the context of host-pathogen interaction studies, dual scRNA-Seq is commonly utilised. ScRNA-Seq involves an extra step which is isolating single cells from tissue samples using techniques like fluorescence-activated cell sorting (FACS), micro-dissection and droplet-based methods instead of bulk sequencing various cell populations [218]. While dual RNA-Seq provides insight about the bigger picture, dual scRNA-Seq can elucidate the smaller scale interactions that sum up to produce the host outcome during infection [219].

It is common for bacteria to have distinct co-existing subpopulations due to their dynamic adaptability. This heterogeneity can lead to phenotypic variations in infection and scRNA-Seq is capable of characterising these variabilities [220]. Avraham et al. [220] examined individual macrophages infected with Salmonella typhimurium and found molecular variations despite what seemed to be identical infections in these cells. They discovered that the type I interferon response pathway is influenced by PhoPQ activity levels in the bacterium. Host cells infected with a bacterium expressing high levels of PhoPQ had an increased type I interferon response. Another similar study also examined bone marrow-derived macrophages exposed to Salmonella with their method called scDual-Seq [221]. From their time-dependent analysis of macrophage single-cell transcriptomes, they found that within infected cells, some had fully induced immune responses while others only had 'partially induced' immune responses. They also found two intracellular classes of Salmonella having unique transcriptional signatures. One of their interesting findings is how the infection progresses from partially induced to fully induced immune responses which also involve changes in Salmonella subpopulations [221].

Meanwhile, scRNA-Seq has also been applied to host-viral interaction studies. In HIV infections, the virus has the ability to persist in latent reservoirs where they are not completely eradicated by treatments like antiretroviral therapy (ART). Golumbeanu et al. [222] used scRNA-Seq to characterise the transcriptomes of latent and reactivated HIV-infected cells. They identified two main subpopulations with one cell cluster being more predisposed to HIV reactivation. Their results provide interesting insights for the identification of potential latency reversing agents and biomarkers for susceptible cells. However, the use of scRNA-Seq in host-pathogen interactions studies are still in its infant stages. Many more questions can be answered using scRNA-Seq such as the mechanism behind selective infections of host cells, antibiotic tolerance of certain bacteria, the switch between active and latent infection in viruses and the list goes on [219].

Furthermore, scRNA-Seq has also played a role in the development of human organoids from stem cells by assessing the similarity between these organoids and primary tissue counterparts [223]. In addition, scRNA-Seq can be used to properly characterise the development and maturation stages of stem cells to specific organ tissue or even used as a blueprint to direct the recreation of actual human organs [224, 225]. Moreover, scRNA-Seq can be used in conjunction with the well-known CRISPR-based gene editing tool to provide confirmation of target gene activation/repression [226]. Advancements in the application of scRNA-Seq in these research areas can provide valuable tools for host-pathogen interaction studies in the future. For instance, the successful creation of human organoids which are highly accurate to real organs can be used as infection models to study disease mechanisms.

Innovations of RNA-Seq methods based on experimental needs have led to its application in various settings. Two of these methods are spatially resolved RNA-Seq known as 'spatialomics' and ribosome-profiling using RNA-Seq to understand the translatome [227]. Spatial information is not provided when using bulk RNA-Seq or scRNA-Seq and this information could be crucial to comprehend cellular processes and how they relate to gene expression. The main concept behind spatialomics is in situ transcriptomics which produce data within tissue sections either using sequencing or imaging [227]. Some of the approaches that have been used in spatial transcriptomics are fluorescent in situ RNA sequencing (FISSEQ) and also a combination of scRNA-Seq data with single molecule fluorescence in situ hybridization method (smFISH) to examine spatial division of genes along liver lobules and investigate gene expression as well as post-transcriptional modifications while preserving spatial information [228, 229]. The smFISH method however had limitations in the number of RNA species that could be imaged at once in single cells. Hence, another method called multiplexed error-robust FISH (MERFISH) was developed which allows thousands of RNA species to be imaged in individual cells with spatial distribution information as well [230]. The use of spatialomics in host pathogen interaction studies shows great promise as many infections by pathogens induce alterations in specific subcellular compartments [231]. Understanding both temporal and spatial changes that occur during the course of an infection can improve our comprehension of host-pathogen interplay. As for ribosome-profiling, the highly regulated process of mRNA translation by ribosomes inspired this translatome-based analysis with an assumption that protein synthesis is proportional to the density of mRNA ribosomes [227]. By sequencing the ribosome-protected mRNAs, studies have gained insight on translational control in yeast, codon usage biases and unannotated translational events [232–234]. Ribosome profiling coupled with RNA-Seq has been carried out as well to study infections by pathogens like *Toxoplasma gondii* and the vaccinia virus. Holmes et al. [235] found open reading frames that may be involved in selective stress-induced translation of parasitic mRNA while Dai et al. [236] found that mRNAs involved in cellular energy production were increased which supported vaccinia virus replication. The applications of RNA-Seq and its combinations with existing methods are increasingly being advanced and modified to suit specific experimental needs.

7. Challenges in RNA-Seq

The rapid surge of RNA-Seq technology has led to many new discoveries and is currently the go-to method for transcriptomic analysis. Although significant advancements have resulted from the use of RNA-Seq, it is still continuously evolving with many aspects that need to be improved. The drawbacks of short-read sequencing platforms as mentioned before have been mostly solved with the advent of long-read technology. While long-read technology has its own strengths, analysing long-read datasets still poses a challenge. Aside from lower accuracies per read compared to short-read platforms, most of the long-read transcriptomic tools do not take into account factors like coverage bias and high error rates [237]. Several studies have found beneficial effects of combining short- and long-read technologies, however integrating different tools are often laborious hence it still needs to be improved [238, 239]. There are certain challenges faced with library preparations as well. In this process, cDNA is generated from fragmented RNAs followed by adapter ligation, amplification and finally sequencing. Linsen et al. [240] compared three different library preparation methods and found that each method had large differences in the frequency of miRNAs captured. Other biases include PCR amplification bias which might be introduced due to variations in template length and base composition during parallel amplification of multiple templates [241, 242]. Yet another issue faced in library preparation is the influence of batch effects. Batch effects may arise from various factors including experimental conditions, quality of reagents, pipetting abilities and also the individual/technician in charge on a particular day [243]. Careful considerations should be made by researchers in order to reduce the effects of these confounding variables.

A recent discovery was the abundance of circular RNAs in various eukaryotic organisms including humans [244]. Previous RNA-Seq protocols were mostly biased against circular RNAs (circRNAs) whereby the poly (A) enrichment step would efficiently deplete all circRNAs since they lack poly (A) tails. The development of alternate protocols more suited to non-coding transcripts like rRNA depletion improved detection of circRNAs. However, these approaches are not entirely efficient for circRNAs and further research is required to improve the detection sensitivity of circRNA and possibly other non-coding RNA transcripts [245]. There are several technical challenges associated with scRNA-Seq as well. With regard to host-bacterial studies, the bacterial lysing protocols employed, whether physical or chemical, are not very compatible with further downstream steps in RNA-Seq like amplification and library preparation. These steps also do not preserve the RNA effectively. Another problem is the accurate identification of minority transcripts in bacteria. ScRNA-Seq protocols commonly employ poly (A) enriching strategies which are useful for eukaryotes however, prokaryotic mRNAs are not poly-adenylated. Analysis of non-polyadenylated RNAs have been attempted however, they involve complex and specialised protocols which need to be simplified [218, 246]. This problem is also faced when analysing viral infections in host cells because certain viruses like dengue virus and hepatitis C virus have non-polyadenylated mRNAs. There needs to be a more optimum procedure to accurately quantify bacterial and viral transcripts. Furthermore, scRNA-Seq examines individual cells leading to very low input material. This results in high levels of technical noise which can be confused with biological variability [247]. A few statistical models have been proposed which are capable of quantifying this technical noise but additional research is required to assess the validity of these models [247, 248].

The development of more complex tools for RNA-Seq analysis are quite possible and challenges may arise in the comprehension or use of such approaches. Efforts should be made to increase the practicality of approaches to avoid methods that are only manageable for those with very high expertise. While many tools exist for the analysis of RNA-Seq data, they seem to be more than we can handle. There are a multitude of pipelines incorporating many different tools with multiple versions and licences [249]. This is a major challenge especially in the context of translating RNA-Seq into clinic. Bringing a laboratory test into clinic involves an important step that is demonstration of analytical validity. One aspect of analytical validity is accuracy that is commonly measured by comparing obtained values to a reference standard [249]. The development of a reference standard especially for NGS data can reduce method- and platform-specific biases [250]. One of the first reference standards that existed for RNA-Seq was developed by the External RNA Controls Consortium (ERCC) using synthetic RNA spike-in controls [251]. Other projects like the Sequencing Quality Control (SEQC) [252], Association of Biomolecular Resource Facilities (ABRF) [253] and GEUVADIS [254] carried out extensive studies investigating the accuracy of RNA-Seq data across many platforms, protocols and laboratory sites, providing a guide for other researchers. The continuous technological advancements occurring in the field of sequencing technologies have to be accompanied by more reference standards [250]. The constant development and assessment of reference standards are required to reduce the variability that arises from the emergence of numerous tools. Conquering this challenge will also allow improved translation of RNA-Seq into clinic and ensure the smooth transition of NGS technologies into clinical settings.

8. Summary

RNA-Seq has revolutionised the approach taken by researchers in exploring host-pathogen interactions. From scRNA-Seq to bulk RNA-Seq, the vast amount of information derived from these studies provide novel insights into the exact mechanisms of disease and host counter- reactions in combating the disease. RNA-Seq has allowed us to examine the mechanisms of gene expression, differentially expressed genes in development or disease, alternative splicing events, gene fusion events, transcriptional regulation and many more. The use of dual RNA-Seq has changed our current perspectives of host-pathogen interactions. It is clear that systems-level alterations are induced by infection all the way from immune responses to metabolic processes. These studies are laying the foundation for more complex interrogations of our immune system and eventually its translation into clinical settings. Other creative innovations to RNA-Seq are also bound to occur as long as the determination to answer biological questions are present. The use of spatialomics seems very promising as it allows the known transcripts to be assessed while preserving the three dimensional suuround-ing of the tissue. This has major implications especially in studies investigating the influence of cellular architecture on infection progression. Single-cell RNA-Seq is also slowly gaining momentum in the field of host-pathogen interaction studies namely due to its ability to elucidate pathogen subpopulations. This is a key factor that will provide further information about their pathogenesis, host cell susceptibility and potential targeted treatment strategies. The current discrepancies and biases that exist within RNA-Seq protocols are challenges that need to be met in order to ensure its upward trajectory. The next few years will be a period of concurrent growth for RNA-Seq technology and biomedical research. A new biological discovery phase has just begun and RNA-Seq has proved to be a valuable tool to guide us through this phase.

Author details

Sudhesh Dev Sareshma[1,2] and Bhassu Subha[1,2*]

1 Genetics and Molecular Biology, Institute of Biological Sciences, Faculty of Science, University Malaya, Kuala Lumpur, Malaysia

2 Center of Biotechnology for Agriculture, University of Malaya, Kuala Lumpur, Malaysia

*Address all correspondence to: subhabhassu@um.edu.my

References

[1] Denoeud F, Aury JM, Da Silva C, Noel B, Rogier O, Delledonne M, et al. Annotating genomes with massive-scale RNA sequencing. Genome biology. 2008;9(12):R175.

[2] Ren S, Peng Z, Mao J-H, Yu Y, Yin C, Gao X, et al. RNA-seq analysis of prostate cancer in the Chinese population identifies recurrent gene fusions, cancer-associated long noncoding RNAs and aberrant alternative splicings. Cell Research. 2012;22(5):806-21.

[3] Gaidatzis D, Burger L, Florescu M, Stadler MB. Analysis of intronic and exonic reads in RNA-seq data characterizes transcriptional and post-transcriptional regulation. Nature Biotechnology. 2015;33(7):722-9.

[4] Pareek CS, Błaszczyk P, Dziuba P, Czarnik U, Fraser L, Sobiech P, et al. Single nucleotide polymorphism discovery in bovine liver using RNA-seq technology. PLOS ONE. 2017;12(2):e0172687.

[5] Zhao H, Chen M, Tellgren-Roth C, Pettersson U. Fluctuating expression of microRNAs in adenovirus infected cells. Virology. 2015;478:99-111.

[6] Rao R, Bing Zhu Y, Alinejad T, Tiruvayipati S, Lin Thong K, Wang J, et al. RNA-seq analysis of *Macrobrachium rosenbergii* hepatopancreas in response to *Vibrio parahaemolyticus* infection. Gut Pathogens. 2015;7(1):6.

[7] Westermann AJ, Barquist L, Vogel J. Resolving host–pathogen interactions by dual RNA-seq. PLOS Pathogens. 2017;13(2):e1006033.

[8] Nagalakshmi U, Wang Z, Waern K, Shou C, Raha D, Gerstein M, et al. The transcriptional landscape of the yeast genome defined by RNA sequencing. Science (New York, NY). 2008;320(5881):1344-9.

[9] Bainbridge MN, Warren RL, Hirst M, Romanuik T, Zeng T, Go A, et al. Analysis of the prostate cancer cell line LNCaP transcriptome using a sequencing-by-synthesis approach. BMC genomics. 2006;7:246.

[10] Cheung F, Haas BJ, Goldberg SM, May GD, Xiao Y, Town CD. Sequencing *Medicago truncatula* expressed sequenced tags using 454 Life Sciences technology. BMC genomics. 2006;7:272.

[11] Emrich SJ, Barbazuk WB, Li L, Schnable PS. Gene discovery and annotation using LCM-454 transcriptome sequencing. Genome research. 2007;17(1):69-73.

[12] Barbazuk WB, Emrich SJ, Chen HD, Li L, Schnable PS. SNP discovery via 454 transcriptome sequencing. The Plant journal : for cell and molecular biology. 2007;51(5):910-8.

[13] Weber AP, Weber KL, Carr K, Wilkerson C, Ohlrogge JB. Sampling the Arabidopsis transcriptome with massively parallel pyrosequencing. Plant physiology. 2007;144(1):32-42.

[14] Barba M, Czosnek H, Hadidi A. Historical perspective, development and applications of next-generation sequencing in plant virology. Viruses. 2014;6(1):106-36.

[15] Wang Z, Gerstein M, Snyder M. RNA-Seq: a revolutionary tool for transcriptomics. Nature reviews Genetics. 2009;10(1):57-63.

[16] Bumgarner R. Overview of DNA microarrays: types, applications, and their future. Current protocols in molecular biology. 2013;Chapter 22:Unit-22.1.

[17] Govindarajan R, Duraiyan J, Kaliyappan K, Palanisamy M. Microarray and its applications. Journal of pharmacy & bioallied sciences. 2012;4(Suppl 2):S310-2.

[18] Fu X, Fu N, Guo S, Yan Z, Xu Y, Hu H, et al. Estimating accuracy of RNA-Seq and microarrays with proteomics. BMC genomics. 2009;10:161.

[19] Choudhuri S. Chapter 3 - Genomic Technologies**The opinions expressed in this chapter are the author's own and they do not necessarily reflect the opinions of the FDA, the DHHS, or the Federal Government. In: Choudhuri S, editor. Bioinformatics for Beginners. Oxford: Academic Press; 2014. p. 55-72.

[20] Held GA, Grinstein G, Tu Y. Relationship between gene expression and observed intensities in DNA microarrays--a modeling study. Nucleic acids research. 2006;34(9):e70.

[21] Okoniewski MJ, Miller CJ. Hybridization interactions between probesets in short oligo microarrays lead to spurious correlations. BMC bioinformatics. 2006;7:276.

[22] Nagaraj SH, Gasser RB, Ranganathan S. A hitchhiker's guide to expressed sequence tag (EST) analysis. Briefings in bioinformatics. 2007;8(1):6-21.

[23] Parkinson J, Blaxter M. Expressed sequence tags: analysis and annotation. Methods in molecular biology (Clifton, NJ). 2004;270:93-126.

[24] Cai J, Shin S, Wright L, Liu Y, Zhou D, Xue H, et al. Massively parallel signature sequencing profiling of fetal human neural precursor cells. Stem cells and development. 2006;15(2):232-44.

[25] Chu TJ, Peters DG. Serial analysis of the vascular endothelial transcriptome under static and shear stress conditions. Physiological Genomics. 2008;34(2):185-92.

[26] Reinartz J, Bruyns E, Lin JZ, Burcham T, Brenner S, Bowen B, et al. Massively parallel signature sequencing (MPSS) as a tool for in-depth quantitative gene expression profiling in all organisms. Briefings in functional genomics & proteomics. 2002;1(1):95-104.

[27] Kodzius R, Kojima M, Nishiyori H, Nakamura M, Fukuda S, Tagami M, et al. CAGE: cap analysis of gene expression. Nature methods. 2006;3(3):211-22.

[28] Fryer RM, Randall J, Yoshida T, Hsiao LL, Blumenstock J, Jensen KE, et al. Global analysis of gene expression: methods, interpretation, and pitfalls. Experimental nephrology. 2002;10(2):64-74.

[29] Zhao X, Valen E, Parker BJ, Sandelin A. Systematic clustering of transcription start site landscapes. PLoS One. 2011;6(8):e23409.

[30] Slatko BE, Gardner AF, Ausubel FM. Overview of Next-Generation Sequencing Technologies. Current protocols in molecular biology. 2018;122(1):e59-e.

[31] Lister R, O'Malley RC, Tonti-Filippini J, Gregory BD, Berry CC, Millar AH, et al. Highly integrated single-base resolution maps of the epigenome in Arabidopsis. Cell. 2008;133(3):523-36.

[32] Wilhelm BT, Marguerat S, Watt S, Schubert F, Wood V, Goodhead I, et al. Dynamic repertoire of a eukaryotic transcriptome surveyed at single-nucleotide resolution. Nature. 2008;453(7199):1239-43.

[33] Morin R, Bainbridge M, Fejes A, Hirst M, Krzywinski M, Pugh T, et al. Profiling the HeLa S3 transcriptome

using randomly primed cDNA and massively parallel short-read sequencing. BioTechniques. 2008;45(1):81-94.

[34] Mortazavi A, Williams BA, McCue K, Schaeffer L, Wold B. Mapping and quantifying mammalian transcriptomes by RNA-Seq. Nature methods. 2008;5(7):621-8.

[35] Cloonan N, Forrest AR, Kolle G, Gardiner BB, Faulkner GJ, Brown MK, et al. Stem cell transcriptome profiling via massive-scale mRNA sequencing. Nature methods. 2008;5(7):613-9.

[36] Tirosh I, Venteicher AS, Hebert C, Escalante LE, Patel AP, Yizhak K, et al. Single-cell RNA-seq supports a developmental hierarchy in human oligodendroglioma. Nature. 2016;539(7628):309-13.

[37] Kremer LS, Bader DM, Mertes C, Kopajtich R, Pichler G, Iuso A, et al. Genetic diagnosis of Mendelian disorders via RNA sequencing. Nature communications. 2017;8:15824.

[38] MacParland SA, Liu JC, Ma XZ, Innes BT, Bartczak AM, Gage BK, et al. Single cell RNA sequencing of human liver reveals distinct intrahepatic macrophage populations. Nature communications. 2018;9(1):4383.

[39] James KL, de Silva TI, Brown K, Whittle H, Taylor S, McVean G, et al. Low-Bias RNA Sequencing of the HIV-2 Genome from Blood Plasma. Journal of virology. 2019;93(1).

[40] Bai Y, Wang D, Li W, Huang Y, Ye X, Waite J, et al. Evaluation of the capacities of mouse TCR profiling from short read RNA-seq data. PLoS One. 2018;13(11):e0207020.

[41] Townsley BT, Covington MF, Ichihashi Y, Zumstein K, Sinha NR. BrAD-seq: Breath Adapter Directional sequencing: a streamlined, ultra-simple and fast library preparation protocol for strand specific mRNA library construction. Frontiers in plant science. 2015;6:366.

[42] Langevin SA, Bent ZW, Solberg OD, Curtis DJ, Lane PD, Williams KP, et al. Peregrine: A rapid and unbiased method to produce strand-specific RNA-Seq libraries from small quantities of starting material. RNA biology. 2013;10(4):502-15.

[43] Schroeder A, Mueller O, Stocker S, Salowsky R, Leiber M, Gassmann M, et al. The RIN: an RNA integrity number for assigning integrity values to RNA measurements. BMC molecular biology. 2006;7:3.

[44] Slomovic S, Laufer D, Geiger D, Schuster G. Polyadenylation of ribosomal RNA in human cells. Nucleic acids research. 2006;34(10):2966-75.

[45] Hrdlickova R, Toloue M, Tian B. RNA-Seq methods for transcriptome analysis. Wiley interdisciplinary reviews RNA. 2017;8(1).

[46] Ingolia NT, Brar GA, Rouskin S, McGeachy AM, Weissman JS. The ribosome profiling strategy for monitoring translation in vivo by deep sequencing of ribosome-protected mRNA fragments. Nature protocols. 2012;7(8):1534-50.

[47] Archer SK, Shirokikh NE, Preiss T. Selective and flexible depletion of problematic sequences from RNA-seq libraries at the cDNA stage. BMC genomics. 2014;15(1):401.

[48] Armour CD, Castle JC, Chen R, Babak T, Loerch P, Jackson S, et al. Digital transcriptome profiling using selective hexamer priming for cDNA synthesis. Nature methods. 2009;6(9):647-9.

[49] Zhao S, Zhang Y, Gamini R, Zhang B, von Schack D. Evaluation of

two main RNA-seq approaches for gene quantification in clinical RNA sequencing: polyA+ selection versus rRNA depletion. Scientific reports. 2018;8(1):4781.

[50] Picelli S, Björklund AK, Reinius B, Sagasser S, Winberg G, Sandberg R. Tn5 transposase and tagmentation procedures for massively scaled sequencing projects. Genome research. 2014;24(12):2033-40.

[51] Kulski JK. Next-generation sequencing—an overview of the history, tools, and "Omic" applications. 2016:3-60.

[52] Guo J, Xu N, Li Z, Zhang S, Wu J, Kim DH, et al. Four-color DNA sequencing with 3'-O-modified nucleotide reversible terminators and chemically cleavable fluorescent dideoxynucleotides. Proceedings of the National Academy of Sciences of the United States of America. 2008;105(27):9145-50.

[53] Specifications for the NextSeq 1000 and NextSeq 2000 Systems n.d. [Available from: https://www.illumina.com/systems/sequencing-platforms/nextseq-1000-2000/specifications.html.

[54] Schadt EE, Turner S, Kasarskis A. A window into third-generation sequencing. Human molecular genetics. 2010;19(R2):R227-40.

[55] Introducing The Sequel IIe System - Sequencing Evolved n.d. [Available from: https://www.pacb.com/products-and-services/sequel-system/latest-system-release/.

[56] Kasianowicz JJ, Brandin E, Branton D, Deamer DW. Characterization of individual polynucleotide molecules using a membrane channel. Proceedings of the National Academy of Sciences of the United States of America. 1996;93(24):13770-3.

[57] Reuter JA, Spacek DV, Snyder MP. High-throughput sequencing technologies. Molecular cell. 2015;58(4):586-97.

[58] Oxford Nanopore Technologies n.d. [Available from: https://nanoporetech.com/products.

[59] Metzker ML. Sequencing technologies - the next generation. Nature reviews Genetics. 2010;11(1):31-46.

[60] Shendure J, Ji H. Next-generation DNA sequencing. Nat Biotechnol. 2008;26(10):1135-45.

[61] Rothberg JM, Hinz W, Rearick TM, Schultz J, Mileski W, Davey M, et al. An integrated semiconductor device enabling non-optical genome sequencing. Nature. 2011;475(7356):348-52.

[62] Thompson JF, Steinmann KE. Single molecule sequencing with a HeliScope genetic analysis system. Current protocols in molecular biology. 2010;Chapter 7:Unit7.10.

[63] Snyder M, Du J, Gerstein M. Personal genome sequencing: current approaches and challenges. Genes & development. 2010;24(5):423-31.

[64] Laver T, Harrison J, O'Neill PA, Moore K, Farbos A, Paszkiewicz K, et al. Assessing the performance of the Oxford Nanopore Technologies MinION. Biomolecular detection and quantification. 2015;3:1-8.

[65] Weirather JL, de Cesare M, Wang Y, Piazza P, Sebastiano V, Wang XJ, et al. Comprehensive comparison of Pacific Biosciences and Oxford Nanopore Technologies and their applications to transcriptome analysis. F1000Research. 2017;6:100.

[66] Petersen LM, Martin IW, Moschetti WE, Kershaw CM,

Tsongalis GJ. Third-Generation Sequencing in the Clinical Laboratory: Exploring the Advantages and Challenges of Nanopore Sequencing. Journal of clinical microbiology. 2019;58(1).

[67] Faster sequencing and data processing n.d. [Available from: https://www.illumina.com/science/technology/next-generation-sequencing/sequencing-technology/2-channel-sbs.html.

[68] Ferrarini M, Moretto M, Ward JA, Šurbanovski N, Stevanović V, Giongo L, et al. An evaluation of the PacBio RS platform for sequencing and de novo assembly of a chloroplast genome. BMC genomics. 2013;14:670.

[69] Teng JLL, Yeung ML, Chan E, Jia L, Lin CH, Huang Y, et al. PacBio But Not Illumina Technology Can Achieve Fast, Accurate and Complete Closure of the High GC, Complex Burkholderia pseudomallei Two-Chromosome Genome. Frontiers in microbiology. 2017;8:1448.

[70] Zhang J, Su L, Wang Y, Deng S. Improved High-Throughput Sequencing of the Human Oral Microbiome: From Illumina to PacBio. The Canadian journal of infectious diseases & medical microbiology = Journal canadien des maladies infectieuses et de la microbiologie medicale. 2020;2020:6678872.

[71] Greig DR, Jenkins C, Gharbia S, Dallman TJ. Comparison of single-nucleotide variants identified by Illumina and Oxford Nanopore technologies in the context of a potential outbreak of Shiga toxin-producing *Escherichia coli*. GigaScience. 2019;8(8).

[72] Quail MA, Smith M, Coupland P, Otto TD, Harris SR, Connor TR, et al. A tale of three next generation sequencing platforms: comparison of Ion Torrent, Pacific Biosciences and Illumina MiSeq sequencers. BMC genomics. 2012;13:341.

[73] Lahens NF, Ricciotti E, Smirnova O, Toorens E, Kim EJ, Baruzzo G, et al. A comparison of Illumina and Ion Torrent sequencing platforms in the context of differential gene expression. BMC genomics. 2017;18(1):602.

[74] Suzuki S, Ranade S, Osaki K, Ito S, Shigenari A, Ohnuki Y, et al. Reference Grade Characterization of Polymorphisms in Full-Length HLA Class I and II Genes With Short-Read Sequencing on the ION PGM System and Long-Reads Generated by Single Molecule, Real-Time Sequencing on the PacBio Platform. Frontiers in immunology. 2018;9:2294.

[75] Tan MH, Austin CM, Hammer MP, Lee YP, Croft LJ, Gan HM. Finding Nemo: hybrid assembly with Oxford Nanopore and Illumina reads greatly improves the clownfish (*Amphiprion ocellaris*) genome assembly. GigaScience. 2018;7(3):1-6.

[76] Guerrero-Sanchez VM, Maldonado-Alconada AM, Amil-Ruiz F, Verardi A, Jorrín-Novo JV, Rey MD. Ion Torrent and lllumina, two complementary RNA-seq platforms for constructing the holm oak (*Quercus ilex*) transcriptome. PLoS One. 2019;14(1):e0210356.

[77] Dhar R, Seethy A, Pethusamy K, Singh S, Rohil V, Purkayastha K, et al. De novo assembly of the Indian blue peacock (*Pavo cristatus*) genome using Oxford Nanopore technology and Illumina sequencing. GigaScience. 2019;8(5).

[78] Li W, Li K, Zhang QJ, Zhu T, Zhang Y, Shi C, et al. Improved hybrid de novo genome assembly and annotation of African wild rice, *Oryza longistaminata*, from Illumina and PacBio sequencing reads. The plant genome. 2020;13(1):e20001.

[79] Huang B, Rong H, Ye Y, Ni Z, Xu M, Zhang W, et al. Transcriptomic analysis of flower color variation in the ornamental crabapple (Malus spp.) half-sib family through Illumina and PacBio Sequel sequencing. Plant physiology and biochemistry : PPB. 2020;149:27-35.

[80] Advantages of paired-end and single-read sequencing n.d. [Available from: https://www.illumina.com/science/technology/next-generation-sequencing/plan-experiments/paired-end-vs-single-read.html.

[81] Corley SM, MacKenzie KL, Beverdam A, Roddam LF, Wilkins MR. Differentially expressed genes from RNA-Seq and functional enrichment results are affected by the choice of single-end versus paired-end reads and stranded versus non-stranded protocols. BMC genomics. 2017;18(1):399.

[82] Tavassoly I, Goldfarb J, Iyengar R. Systems biology primer: the basic methods and approaches. Essays in biochemistry. 2018;62(4):487-500.

[83] Kitano H. Systems biology: a brief overview. Science (New York, NY). 2002;295(5560):1662-4.

[84] Dix A, Vlaic S, Guthke R, Linde J. Use of systems biology to decipher host-pathogen interaction networks and predict biomarkers. Clinical microbiology and infection : the official publication of the European Society of Clinical Microbiology and Infectious Diseases. 2016;22(7):600-6.

[85] Cesur MF, Durmuş S. Systems Biology Modeling to Study Pathogen-Host Interactions. Methods in molecular biology (Clifton, NJ). 2018;1734:97-112.

[86] Westermann AJ, Gorski SA, Vogel J. Dual RNA-seq of pathogen and host. Nature reviews Microbiology. 2012;10(9):618-30.

[87] Falkow S, Isberg RR, Portnoy DA. The interaction of bacteria with mammalian cells. Annual review of cell biology. 1992;8:333-63.

[88] Saliba A-E, C Santos S, Vogel J. New RNA-seq approaches for the study of bacterial pathogens. Current Opinion in Microbiology. 2017;35:78-87.

[89] Westermann AJ, Förstner KU, Amman F, Barquist L, Chao Y, Schulte LN, et al. Dual RNA-seq unveils noncoding RNA functions in host-pathogen interactions. Nature. 2016;529(7587):496-501.

[90] Baddal B, Muzzi A, Censini S, Calogero RA, Torricelli G, Guidotti S, et al. Dual RNA-seq of Nontypeable *Haemophilus influenzae* and Host Cell Transcriptomes Reveals Novel Insights into Host-Pathogen Cross Talk. mBio. 2015;6(6):e01765-15.

[91] Aprianto R, Slager J, Holsappel S, Veening JW. Time-resolved dual RNA-seq reveals extensive rewiring of lung epithelial and pneumococcal transcriptomes during early infection. Genome biology. 2016;17(1):198.

[92] Ritchie ND, Evans TJ. Dual RNA-seq in *Streptococcus pneumoniae* Infection Reveals Compartmentalized Neutrophil Responses in Lung and Pleural Space. mSystems. 2019;4(4).

[93] Rienksma RA, Suarez-Diez M, Mollenkopf HJ, Dolganov GM, Dorhoi A, Schoolnik GK, et al. Comprehensive insights into transcriptional adaptation of intracellular mycobacteria by microbe-enriched dual RNA sequencing. BMC genomics. 2015;16(1):34.

[94] Pisu D, Huang L, Grenier JK, Russell DG. Dual RNA-Seq of Mtb-Infected Macrophages In Vivo Reveals Ontologically Distinct Host-Pathogen Interactions. Cell reports. 2020;30(2):335-50.e4.

[95] Montoya DJ, Andrade P, Silva BJA, Teles RMB, Ma F, Bryson B, et al. Dual RNA-Seq of Human Leprosy Lesions Identifies Bacterial Determinants Linked to Host Immune Response. Cell reports. 2019;26(13):3574-85.e3.

[96] Yimthin T, Cliff JM, Phunpang R, Ekchariyawat P, Kaewarpai T, Lee JS, et al. Blood transcriptomics to characterize key biological pathways and identify biomarkers for predicting mortality in melioidosis. Emerging microbes & infections. 2020:1-47.

[97] Whitaker-Dowling P, Youngner JS. VIRUS-HOST CELL INTERACTIONS. Encyclopedia of Virology. 1999:1957-61.

[98] Lisnic VJ, Babic Cac M, Lisnic B, Trsan T, Mefferd A, Das Mukhopadhyay C, et al. Dual analysis of the murine cytomegalovirus and host cell transcriptomes reveal new aspects of the virus-host cell interface. PLoS Pathog. 2013;9(9):e1003611.

[99] Park SJ, Kumar M, Kwon HI, Seong RK, Han K, Song JM, et al. Dynamic changes in host gene expression associated with H5N8 avian influenza virus infection in mice. Scientific reports. 2015;5:16512.

[100] Jones M, Dry IR, Frampton D, Singh M, Kanda RK, Yee MB, et al. RNA-seq analysis of host and viral gene expression highlights interaction between varicella zoster virus and keratinocyte differentiation. PLoS Pathog. 2014;10(1):e1003896.

[101] Kozak RA, Fraser RS, Biondi MJ, Majer A, Medina SJ, Griffin BD, et al. Dual RNA-Seq characterization of host and pathogen gene expression in liver cells infected with Crimean-Congo Hemorrhagic Fever Virus. PLoS neglected tropical diseases. 2020;14(4):e0008105.

[102] Fabozzi G, Oler AJ, Liu P, Chen Y, Mindaye S, Dolan MA, et al. Strand-Specific Dual RNA Sequencing of Bronchial Epithelial Cells Infected with Influenza A/H3N2 Viruses Reveals Splicing of Gene Segment 6 and Novel Host-Virus Interactions. Journal of virology. 2018;92(17).

[103] Michlmayr D, Kim EY, Rahman AH, Raghunathan R, Kim-Schulze S, Che Y, et al. Comprehensive Immunoprofiling of Pediatric Zika Reveals Key Role for Monocytes in the Acute Phase and No Effect of Prior Dengue Virus Infection. Cell reports. 2020;31(4):107569.

[104] Wesolowska-Andersen A, Everman JL, Davidson R, Rios C, Herrin R, Eng C, et al. Dual RNA-seq reveals viral infections in asthmatic children without respiratory illness which are associated with changes in the airway transcriptome. Genome biology. 2017;18(1):12.

[105] Sridhar S, To KK, Chan JF, Lau SK, Woo PC, Yuen KY. A systematic approach to novel virus discovery in emerging infectious disease outbreaks. The Journal of molecular diagnostics : JMD. 2015;17(3):230-41.

[106] Chen L, Liu W, Zhang Q, Xu K, Ye G, Wu W, et al. RNA based mNGS approach identifies a novel human coronavirus from two individual pneumonia cases in 2019 Wuhan outbreak. Emerging microbes & infections. 2020;9(1):313-9.

[107] Cao M, Zhang S, Li M, Liu Y, Dong P, Li S, et al. Discovery of Four Novel Viruses Associated with Flower Yellowing Disease of Green Sichuan Pepper (Zanthoxylum Armatum) by Virome Analysis. Viruses. 2019;11(8).

[108] Wright AA, Cross AR, Harper SJ. A bushel of viruses: Identification of seventeen novel putative viruses by RNA-seq in six apple trees. PLoS One. 2020;15(1):e0227669.

[109] Schmit JP, Mueller GM. An estimate of the lower limit of global fungal diversity. Biodiversity and Conservation. 2007;16(1):99-111.

[110] Horn F, Heinekamp T, Kniemeyer O, Pollmächer J, Valiante V, Brakhage AA. Systems biology of fungal infection. Frontiers in microbiology. 2012;3:108.

[111] McCormick A, Heesemann L, Wagener J, Marcos V, Hartl D, Loeffler J, et al. NETs formed by human neutrophils inhibit growth of the pathogenic mold Aspergillus fumigatus. Microbes and infection. 2010;12(12-13):928-36.

[112] Moalli F, Doni A, Deban L, Zelante T, Zagarella S, Bottazzi B, et al. Role of complement and Fc{gamma} receptors in the protective activity of the long pentraxin PTX3 against Aspergillus fumigatus. Blood. 2010;116(24):5170-80.

[113] Thywißen A, Heinekamp T, Dahse HM, Schmaler-Ripcke J, Nietzsche S, Zipfel PF, et al. Conidial Dihydroxynaphthalene Melanin of the Human Pathogenic Fungus Aspergillus fumigatus Interferes with the Host Endocytosis Pathway. Frontiers in microbiology. 2011;2:96.

[114] Rizzetto L, Cavalieri D. Friend or foe: using systems biology to elucidate interactions between fungi and their hosts. Trends in microbiology. 2011;19(10):509-15.

[115] Bruno VM, Wang Z, Marjani SL, Euskirchen GM, Martin J, Sherlock G, et al. Comprehensive annotation of the transcriptome of the human fungal pathogen *Candida albicans* using RNA-seq. Genome research. 2010;20(10):1451-8.

[116] Linde J, Duggan S, Weber M, Horn F, Sieber P, Hellwig D, et al. Defining the transcriptomic landscape of *Candida glabrata* by RNA-Seq. Nucleic acids research. 2015;43(3):1392-406.

[117] Tierney L, Linde J, Müller S, Brunke S, Molina J, Hube B, et al. An Interspecies Regulatory Network Inferred from Simultaneous RNA-seq of *Candida albicans* Invading Innate Immune Cells. 2012;3(85).

[118] Sieber P, Voigt K, Kämmer P, Brunke S, Schuster S, Linde J. Comparative Study on Alternative Splicing in Human Fungal Pathogens Suggests Its Involvement During Host Invasion. Frontiers in microbiology. 2018;9:2313.

[119] Zhang Q, Zhang J, Gong M, Pan R, Liu Y, Tao L, et al. Transcriptome Analysis of the Gene Expression Profiles Associated with Fungal Keratitis in Mice Based on RNA-Seq. Investigative ophthalmology & visual science. 2020;61(6):32.

[120] Petrucelli MF, Peronni K, Sanches PR, Komoto TT, Matsuda JB, Silva Junior WAD, et al. Dual RNA-Seq Analysis of Trichophyton rubrum and HaCat Keratinocyte Co-Culture Highlights Important Genes for Fungal-Host Interaction. Genes. 2018;9(7).

[121] Lass-Flörl C, Mayr A. Human protothecosis. Clinical microbiology reviews. 2007;20(2):230-42.

[122] Geschwind MD. Prion Diseases. Continuum (Minneapolis, Minn). 2015;21(6 Neuroinfectious Disease):1612-38.

[123] Mitchell PD. The origins of human parasites: Exploring the evidence for endoparasitism throughout human evolution. International journal of paleopathology. 2013;3(3):191-8.

[124] Blasco-Costa I, Poulin R. Parasite life-cycle studies: a plea to resurrect an

old parasitological tradition. Journal of helminthology. 2017;91(6):647-56.

[125] Ngara M, Palmkvist M, Sagasser S, Hjelmqvist D, Björklund Å K, Wahlgren M, et al. Exploring parasite heterogeneity using single-cell RNA-seq reveals a gene signature among sexual stage *Plasmodium falciparum* parasites. Experimental cell research. 2018;371(1):130-8.

[126] Greif G, Ponce de Leon M, Lamolle G, Rodriguez M, Piñeyro D, Tavares-Marques LM, et al. Transcriptome analysis of the bloodstream stage from the parasite Trypanosoma vivax. BMC genomics. 2013;14:149.

[127] Choi YJ, Aliota MT, Mayhew GF, Erickson SM, Christensen BM. Dual RNA-seq of parasite and host reveals gene expression dynamics during filarial worm-mosquito interactions. PLoS neglected tropical diseases. 2014;8(5):e2905.

[128] Foth BJ, Tsai IJ, Reid AJ, Bancroft AJ, Nichol S, Tracey A, et al. Whipworm genome and dual-species transcriptome analyses provide molecular insights into an intimate host-parasite interaction. Nature genetics. 2014;46(7):693-700.

[129] Anderson L, Amaral MS, Beckedorff F, Silva LF, Dazzani B, Oliveira KC, et al. *Schistosoma mansoni* Egg, Adult Male and Female Comparative Gene Expression Analysis and Identification of Novel Genes by RNA-Seq. PLoS neglected tropical diseases. 2015;9(12):e0004334.

[130] Pittman KJ, Aliota MT, Knoll LJ. Dual transcriptional profiling of mice and Toxoplasma gondii during acute and chronic infection. BMC genomics. 2014;15(1):806.

[131] Soto C, Satani N. The intricate mechanisms of neurodegeneration in prion diseases. Trends in molecular medicine. 2011;17(1):14-24.

[132] Bellingham SA, Coleman BM, Hill AF. Small RNA deep sequencing reveals a distinct miRNA signature released in exosomes from prion-infected neuronal cells. Nucleic acids research. 2012;40(21):10937-49.

[133] Carroll JA, Race B, Williams K, Striebel J, Chesebro B. RNA-seq and network analysis reveal unique glial gene expression signatures during prion infection. Molecular brain. 2020;13(1):71.

[134] Thackray AM, Lam B, Shahira Binti Ab Razak A, Yeo G, Bujdoso R. Transcriptional signature of prion-induced neurotoxicity in a Drosophila model of transmissible mammalian prion disease. The Biochemical journal. 2020;477(4):833-52.

[135] Bakuła Z, Gromadka R, Gawor J, Siedlecki P, Pomorski JJ, Maciszewski K, et al. Sequencing and Analysis of the Complete Organellar Genomes of Prototheca wickerhamii. Frontiers in plant science. 2020;11:1296.

[136] Zeng X, Kudinha T, Kong F, Zhang QQ. Comparative Genome and Transcriptome Study of the Gene Expression Difference Between Pathogenic and Environmental Strains of Prototheca zopfii. Frontiers in microbiology. 2019;10:443.

[137] Vlasova-St. Louis I, Chang CC, Shahid S, French MA, Bohjanen PR. Transcriptomic Predictors of Paradoxical Cryptococcosis-Associated Immune Reconstitution Inflammatory Syndrome. Open Forum Infectious Diseases. 2018;5(7).

[138] Seelbinder B, Wallstabe J, Marischen L, Weiss E, Wurster S, Page L, et al. Triple RNA-Seq Reveals Synergy in a Human Virus-Fungus Co-infection Model. Cell reports. 2020;33(7):108389.

[139] Conesa A, Madrigal P, Tarazona S, Gomez-Cabrero D, Cervera A, McPherson A, et al. A survey of best practices for RNA-seq data analysis. Genome biology. 2016;17:13.

[140] Alkhateeb A, Rueda L. Zseq: An Approach for Preprocessing Next-Generation Sequencing Data. Journal of computational biology : a journal of computational molecular cell biology. 2017;24(8):746-55.

[141] Zhao S, Zhang B, Zhang Y, Gordon W, Du S, Paradis T, et al. Bioinformatics for RNA-seq data analysis. 2016:125-49.

[142] Andrews S. FastQC: a quality control tool for high throughput sequence data. Babraham Bioinformatics, Babraham Institute, Cambridge, United Kingdom; 2010.

[143] Wang L, Wang S, Li W. RSeQC: quality control of RNA-seq experiments. Bioinformatics (Oxford, England). 2012;28(16):2184-5.

[144] Dai M, Thompson RC, Maher C, Contreras-Galindo R, Kaplan MH, Markovitz DM, et al. NGSQC: cross-platform quality analysis pipeline for deep sequencing data. BMC genomics. 2010;11 Suppl 4(Suppl 4):S7.

[145] Bolger AM, Lohse M, Usadel B. Trimmomatic: a flexible trimmer for Illumina sequence data. Bioinformatics (Oxford, England). 2014;30(15):2114-20.

[146] Martin MJEj. Cutadapt removes adapter sequences from high-throughput sequencing reads. 2011;17(1):10-2.

[147] Liu X, Yan Z, Wu C, Yang Y, Li X, Zhang G. FastProNGS: fast preprocessing of next-generation sequencing reads. BMC bioinformatics. 2019;20(1):345.

[148] Martínez-Alcántara A, Ballesteros E, Feng C, Rojas M, Koshinsky H, Fofanov VY, et al. PIQA: pipeline for Illumina G1 genome analyzer data quality assessment. Bioinformatics (Oxford, England). 2009;25(18):2438-9.

[149] Pérez-Rubio P, Lottaz C, Engelmann JC. FastqPuri: high-performance preprocessing of RNA-seq data. BMC bioinformatics. 2019;20(1):226.

[150] Zhou Q, Su X, Jing G, Chen S, Ning K. RNA-QC-chain: comprehensive and fast quality control for RNA-Seq data. BMC genomics. 2018;19(1):144.

[151] Chen S, Zhou Y, Chen Y, Gu J. fastp: an ultra-fast all-in-one FASTQ preprocessor. Bioinformatics (Oxford, England). 2018;34(17):i884-i90.

[152] Dobin A, Davis CA, Schlesinger F, Drenkow J, Zaleski C, Jha S, et al. STAR: ultrafast universal RNA-seq aligner. Bioinformatics (Oxford, England). 2013;29(1):15-21.

[153] Kim D, Pertea G, Trapnell C, Pimentel H, Kelley R, Salzberg SL. TopHat2: accurate alignment of transcriptomes in the presence of insertions, deletions and gene fusions. Genome biology. 2013;14(4):R36.

[154] Wang K, Singh D, Zeng Z, Coleman SJ, Huang Y, Savich GL, et al. MapSplice: accurate mapping of RNA-seq reads for splice junction discovery. Nucleic acids research. 2010;38(18):e178.

[155] Langmead B, Salzberg SL. Fast gapped-read alignment with Bowtie 2. Nature methods. 2012;9(4):357-9.

[156] Boratyn GM, Thierry-Mieg J, Thierry-Mieg D, Busby B, Madden TL. Magic-BLAST, an accurate RNA-seq aligner for long and short reads. BMC bioinformatics. 2019;20(1):405.

[157] Baruzzo G, Hayer KE, Kim EJ, Di Camillo B, FitzGerald GA, Grant GR. Simulation-based comprehensive benchmarking of RNA-seq aligners. Nature methods. 2017;14(2):135-9.

[158] Schaarschmidt S, Fischer A, Zuther E, Hincha DK. Evaluation of Seven Different RNA-Seq Alignment Tools Based on Experimental Data from the Model Plant *Arabidopsis thaliana*. International journal of molecular sciences. 2020;21(5).

[159] Engström PG, Steijger T, Sipos B, Grant GR, Kahles A, Rätsch G, et al. Systematic evaluation of spliced alignment programs for RNA-seq data. Nature methods. 2013;10(12):1185-91.

[160] Roberts A, Pimentel H, Trapnell C, Pachter L. Identification of novel transcripts in annotated genomes using RNA-Seq. Bioinformatics (Oxford, England). 2011;27(17):2325-9.

[161] Li JJ, Jiang C-R, Brown JB, Huang H, Bickel PJJPotNAoS. Sparse linear modeling of next-generation mRNA sequencing (RNA-Seq) data for isoform discovery and abundance estimation. 2011;108(50):19867-72.

[162] Mezlini AM, Smith EJ, Fiume M, Buske O, Savich GL, Shah S, et al. iReckon: simultaneous isoform discovery and abundance estimation from RNA-seq data. Genome research. 2013;23(3):519-29.

[163] Pertea M, Pertea GM, Antonescu CM, Chang T-C, Mendell JT, Salzberg SLJNb. StringTie enables improved reconstruction of a transcriptome from RNA-seq reads. 2015;33(3):290-5.

[164] Haas BJ, Papanicolaou A, Yassour M, Grabherr M, Blood PD, Bowden J, et al. De novo transcript sequence reconstruction from RNA-seq using the Trinity platform for reference generation and analysis. Nature protocols. 2013;8(8):1494-512.

[165] Xie Y, Wu G, Tang J, Luo R, Patterson J, Liu S, et al. SOAPdenovo-Trans: de novo transcriptome assembly with short RNA-Seq reads. Bioinformatics (Oxford, England). 2014;30(12):1660-6.

[166] Robertson G, Schein J, Chiu R, Corbett R, Field M, Jackman SD, et al. De novo assembly and analysis of RNA-seq data. Nature methods. 2010;7(11):909-12.

[167] Schulz MH, Zerbino DR, Vingron M, Birney E. Oases: robust de novo RNA-seq assembly across the dynamic range of expression levels. Bioinformatics (Oxford, England). 2012;28(8):1086-92.

[168] Anders S, Pyl PT, Huber WJB. HTSeq—a Python framework to work with high-throughput sequencing data. 2015;31(2):166-9.

[169] Liao Y, Smyth GK, Shi WJB. featureCounts: an efficient general purpose program for assigning sequence reads to genomic features. 2014;30(7):923-30.

[170] Zhang C, Zhang B, Vincent M, Zhao S. Bioinformatics tools for RNA-seq gene and isoform quantification. 2016;3:140.

[171] Li B, Dewey CNJBb. RSEM: accurate transcript quantification from RNA-Seq data with or without a reference genome. 2011;12(1):323.

[172] Roberts A, Pachter LJNm. Streaming fragment assignment for real-time analysis of sequencing experiments. 2013;10(1):71-3.

[173] Nariai N, Kojima K, Mimori T, Sato Y, Kawai Y, Yamaguchi-Kabata Y, et al. TIGAR2: sensitive and accurate

estimation of transcript isoform expression with longer RNA-Seq reads. 2014;15(S10):S5.

[174] Patro R, Duggal G, Love MI, Irizarry RA, Kingsford CJNm. Salmon provides fast and bias-aware quantification of transcript expression. 2017;14(4):417-9.

[175] Bray NL, Pimentel H, Melsted P, Pachter LJNb. Near-optimal probabilistic RNA-seq quantification. 2016;34(5):525-7.

[176] Patro R, Mount SM, Kingsford CJNb. Sailfish enables alignment-free isoform quantification from RNA-seq reads using lightweight algorithms. 2014;32(5):462-4.

[177] Li DJEP. Statistical Methods for RNA Sequencing Data Analysis. 2019:85-99.

[178] Hardcastle TJ, Kelly KA. baySeq: empirical Bayesian methods for identifying differential expression in sequence count data. BMC bioinformatics. 2010;11:422.

[179] Anders S, Huber W. Differential expression analysis for sequence count data. Genome biology. 2010;11(10):R106.

[180] Love MI, Huber W, Anders S. Moderated estimation of fold change and dispersion for RNA-seq data with DESeq2. Genome biology. 2014;15(12):550.

[181] Leng N, Dawson JA, Thomson JA, Ruotti V, Rissman AI, Smits BM, et al. EBSeq: an empirical Bayes hierarchical model for inference in RNA-seq experiments. Bioinformatics (Oxford, England). 2013;29(8):1035-43.

[182] Robinson MD, McCarthy DJ, Smyth GK. edgeR: a Bioconductor package for differential expression analysis of digital gene expression data. Bioinformatics (Oxford, England). 2010;26(1):139-40.

[183] Di Y, Schafer D, Cumbie J, Chang J. NBPSeq: Negative Binomial Models for RNA-Sequencing Data R package version 0.3. 0, URL http://CRAN. R-project. org/package= NBPSeq. 2015.

[184] Li J, Witten DM, Johnstone IM, Tibshirani R. Normalization, testing, and false discovery rate estimation for RNA-sequencing data. Biostatistics. 2012;13(3):523-38.

[185] Auer PL, Doerge RWJSaig, biology m. A two-stage Poisson model for testing RNA-seq data. 2011;10(1).

[186] Trapnell C, Williams BA, Pertea G, Mortazavi A, Kwan G, van Baren MJ, et al. Transcript assembly and quantification by RNA-Seq reveals unannotated transcripts and isoform switching during cell differentiation. Nat Biotechnol. 2010;28(5):511-5.

[187] Trapnell C, Hendrickson DG, Sauvageau M, Goff L, Rinn JL, Pachter L. Differential analysis of gene regulation at transcript resolution with RNA-seq. Nat Biotechnol. 2013;31(1):46-53.

[188] Tarazona S, García-Alcalde F, Dopazo J, Ferrer A, Conesa A. Differential expression in RNA-seq: a matter of depth. Genome research. 2011;21(12):2213-23.

[189] Li J, Tibshirani R. Finding consistent patterns: a nonparametric approach for identifying differential expression in RNA-Seq data. Stat Methods Med Res. 2013;22(5):519-36.

[190] Smyth GK. Linear models and empirical bayes methods for assessing differential expression in microarray experiments. Statistical applications in genetics and molecular biology. 2004;3:Article3.

[191] Law CW, Chen Y, Shi W, Smyth GK. voom: Precision weights unlock linear model analysis tools for RNA-seq read counts. Genome biology. 2014;15(2):R29.

[192] Wang L, Feng Z, Wang X, Wang X, Zhang X. DEGseq: an R package for identifying differentially expressed genes from RNA-seq data. Bioinformatics (Oxford, England). 2010;26(1):136-8.

[193] van de Wiel MA, Neerincx M, Buffart TE, Sie D, Verheul HMW. ShrinkBayes: a versatile R-package for analysis of count-based sequencing data in complex study designs. BMC bioinformatics. 2014;15(1):116.

[194] Davidson NM, Oshlack A. Corset: enabling differential gene expression analysis for de novo assembled transcriptomes. Genome biology. 2014;15(7):410.

[195] Robles JA, Qureshi SE, Stephen SJ, Wilson SR, Burden CJ, Taylor JM. Efficient experimental design and analysis strategies for the detection of differential expression using RNA-Sequencing. BMC genomics. 2012;13:484.

[196] Soneson C, Delorenzi M. A comparison of methods for differential expression analysis of RNA-seq data. BMC bioinformatics. 2013;14:91.

[197] Rapaport F, Khanin R, Liang Y, Pirun M, Krek A, Zumbo P, et al. Comprehensive evaluation of differential gene expression analysis methods for RNA-seq data. Genome biology. 2013;14(9):R95.

[198] Zhang ZH, Jhaveri DJ, Marshall VM, Bauer DC, Edson J, Narayanan RK, et al. A comparative study of techniques for differential expression analysis on RNA-Seq data. PLoS One. 2014;9(8):e103207.

[199] Seyednasrollah F, Laiho A, Elo LL. Comparison of software packages for detecting differential expression in RNA-seq studies. Briefings in bioinformatics. 2015;16(1):59-70.

[200] Rajkumar AP, Qvist P, Lazarus R, Lescai F, Ju J, Nyegaard M, et al. Experimental validation of methods for differential gene expression analysis and sample pooling in RNA-seq. BMC genomics. 2015;16(1):548.

[201] Costa-Silva J, Domingues D, Lopes FM. RNA-Seq differential expression analysis: An extended review and a software tool. PLoS One. 2017;12(12):e0190152.

[202] Mehmood A, Laiho A, Venäläinen MS, McGlinchey AJ, Wang N, Elo LL. Systematic evaluation of differential splicing tools for RNA-seq studies. Briefings in bioinformatics. 2020;21(6):2052-65.

[203] Anders S, Reyes A, Huber W. Detecting differential usage of exons from RNA-seq data. Genome research. 2012;22(10):2008-17.

[204] Hartley SW, Mullikin JC. Detection and visualization of differential splicing in RNA-Seq data with JunctionSeq. Nucleic acids research. 2016;44(15):e127.

[205] Vaquero-Garcia J, Barrera A, Gazzara MR, González-Vallinas J, Lahens NF, Hogenesch JB, et al. A new view of transcriptome complexity and regulation through the lens of local splicing variations. eLife. 2016;5:e11752.

[206] Zhu D, Deng N, Bai C. A generalized dSpliceType framework to detect differential splicing and differential expression events using RNA-Seq. IEEE transactions on nanobioscience. 2015;14(2):192-202.

[207] Trincado JL, Entizne JC, Hysenaj G, Singh B, Skalic M, Elliott DJ, et al. SUPPA2: fast, accurate, and

uncertainty-aware differential splicing analysis across multiple conditions. Genome biology. 2018;19(1):40.

[208] Hu Y, Huang Y, Du Y, Orellana CF, Singh D, Johnson AR, et al. DiffSplice: the genome-wide detection of differential splicing events with RNA-seq. Nucleic acids research. 2013;41(2):e39.

[209] Young MD, Wakefield MJ, Smyth GK, Oshlack A. Gene ontology analysis for RNA-seq: accounting for selection bias. Genome biology. 2010;11(2):R14.

[210] Hänzelmann S, Castelo R, Guinney J. GSVA: gene set variation analysis for microarray and RNA-seq data. BMC bioinformatics. 2013;14:7.

[211] Wang X, Cairns MJ. SeqGSEA: a Bioconductor package for gene set enrichment analysis of RNA-Seq data integrating differential expression and splicing. Bioinformatics (Oxford, England). 2014;30(12):1777-9.

[212] Kanehisa M, Goto S, Kawashima S, Okuno Y, Hattori M. The KEGG resource for deciphering the genome. Nucleic acids research. 2004;32(Database issue):D277-80.

[213] Ashburner M, Ball CA, Blake JA, Botstein D, Butler H, Cherry JM, et al. Gene ontology: tool for the unification of biology. The Gene Ontology Consortium. Nature genetics. 2000;25(1):25-9.

[214] Huber W, Carey VJ, Gentleman R, Anders S, Carlson M, Carvalho BS, et al. Orchestrating high-throughput genomic analysis with Bioconductor. Nature methods. 2015;12(2):115-21.

[215] Choi YJ, Aliota MT, Mayhew GF, Erickson SM, Christensen BM. Dual RNA-seq of Parasite and Host Reveals Gene Expression Dynamics during Filarial Worm–Mosquito Interactions. PLoS neglected tropical diseases. 2014;8(5):e2905.

[216] Liao ZX, Ni Z, Wei XL, Chen L, Li JY, Yu YH, et al. Dual RNA-seq of Xanthomonas oryzae pv. oryzicola infecting rice reveals novel insights into bacterial-plant interaction. PLOS ONE. 2019;14(4):e0215039.

[217] Sun Y, Zhuang Z, Wang X, Huang H, Fu Q, Yan Q. Dual RNA-seq reveals the effect of the flgM gene of Pseudomonas plecoglossicida on the immune response of *Epinephelus coioides*. Fish & shellfish immunology. 2019;87:515-23.

[218] Haque A, Engel J, Teichmann SA, Lönnberg T. A practical guide to single-cell RNA-sequencing for biomedical research and clinical applications. Genome Medicine. 2017;9(1):75.

[219] Penaranda C, Hung DT. Single-Cell RNA Sequencing to Understand Host-Pathogen Interactions. ACS infectious diseases. 2019;5(3):336-44.

[220] Avraham R, Haseley N, Brown D, Penaranda C, Jijon HB, Trombetta JJ, et al. Pathogen Cell-to-Cell Variability Drives Heterogeneity in Host Immune Responses. Cell. 2015;162(6):1309-21.

[221] Avital G, Avraham R, Fan A, Hashimshony T, Hung DT, Yanai I. scDual-Seq: mapping the gene regulatory program of Salmonella infection by host and pathogen single-cell RNA-sequencing. Genome biology. 2017;18(1):200.

[222] Golumbeanu M, Cristinelli S, Rato S, Munoz M, Cavassini M, Beerenwinkel N, et al. Single-Cell RNA-Seq Reveals Transcriptional Heterogeneity in Latent and Reactivated HIV-Infected Cells. Cell reports. 2018;23(4):942-50.

[223] Brazovskaja A, Treutlein B, Camp JG. High-throughput single-cell

transcriptomics on organoids. Current opinion in biotechnology. 2019;55:167-71.

[224] Combes AN, Phipson B, Zappia L, Lawlor KT, Er PX, Oshlack A, et al. High throughput single cell RNA-seq of developing mouse kidney and human kidney organoids reveals a roadmap for recreating the kidney. 2017:235499.

[225] Collin J, Queen R, Zerti D, Dorgau B, Hussain R, Coxhead J, et al. Deconstructing Retinal Organoids: Single Cell RNA-Seq Reveals the Cellular Components of Human Pluripotent Stem Cell-Derived Retina. Stem Cells. 2019;37(5):593-8.

[226] Burgess DJ. Genetic screens: Combining CRISPR perturbations and RNA-seq. Nature reviews Genetics. 2017;18(2):67.

[227] Stark R, Grzelak M, Hadfield J. RNA sequencing: the teenage years. Nature reviews Genetics. 2019;20(11):631-56.

[228] Halpern KB, Shenhav R, Matcovitch-Natan O, Toth B, Lemze D, Golan M, et al. Single-cell spatial reconstruction reveals global division of labour in the mammalian liver. Nature. 2017;542(7641):352-6.

[229] Lee JH, Daugharthy ER, Scheiman J, Kalhor R, Ferrante TC, Terry R, et al. Fluorescent in situ sequencing (FISSEQ) of RNA for gene expression profiling in intact cells and tissues. Nature protocols. 2015;10(3):442-58.

[230] Chen KH, Boettiger AN, Moffitt JR, Wang S, Zhuang X. Spatially resolved, highly multiplexed RNA profiling in single cells. Science (New York, NY). 2015;348(6233):aaa6090.

[231] Jean Beltran PM, Federspiel JD, Sheng X, Cristea IM. Proteomics and integrative omic approaches for understanding host–pathogen interactions and infectious diseases. Molecular Systems Biology. 2017;13(3):922.

[232] Hsu PY, Calviello L, Wu HL, Li FW, Rothfels CJ, Ohler U, et al. Super-resolution ribosome profiling reveals unannotated translation events in Arabidopsis. Proceedings of the National Academy of Sciences of the United States of America. 2016;113(45):E7126-e35.

[233] Ingolia NT, Ghaemmaghami S, Newman JR, Weissman JS. Genome-wide analysis in vivo of translation with nucleotide resolution using ribosome profiling. Science (New York, NY). 2009;324(5924):218-23.

[234] Paulet D, David A, Rivals E. Ribo-seq enlightens codon usage bias. DNA research : an international journal for rapid publication of reports on genes and genomes. 2017;24(3):303-210.

[235] Holmes MJ, Shah P, Wek RC, Sullivan WJ. Simultaneous Ribosome Profiling of Human Host Cells Infected with <span class="named-content genus-species" id="named-content-1">Toxoplasma gondii</span>. mSphere. 2019;4(3):e00292-19.

[236] Dai A, Cao S, Dhungel P, Luan Y, Liu Y, Xie Z, et al. Ribosome Profiling Reveals Translational Upregulation of Cellular Oxidative Phosphorylation mRNAs during Vaccinia Virus-Induced Host Shutoff. Journal of virology. 2017;91(5).

[237] Amarasinghe SL, Su S, Dong X, Zappia L, Ritchie ME, Gouil Q. Opportunities and challenges in long-read sequencing data analysis. Genome biology. 2020;21(1):30.

[238] De Maio N, Shaw LP, Hubbard A, George S, Sanderson ND, Swann J, et al. Comparison of long-read sequencing technologies in the hybrid assembly of

complex bacterial genomes. Microbial genomics. 2019;5(9).

[239] Mahmoud M, Gobet N, Cruz-Dávalos DI, Mounier N, Dessimoz C, Sedlazeck FJ. Structural variant calling: the long and the short of it. Genome biology. 2019;20(1):246.

[240] Linsen SE, de Wit E, Janssens G, Heater S, Chapman L, Parkin RK, et al. Limitations and possibilities of small RNA digital gene expression profiling. Nature methods. 2009;6(7):474-6.

[241] Dabney J, Meyer M. Length and GC-biases during sequencing library amplification: a comparison of various polymerase-buffer systems with ancient and modern DNA sequencing libraries. BioTechniques. 2012;52(2):87-94.

[242] Raabe CA, Tang TH, Brosius J, Rozhdestvensky TS. Biases in small RNA deep sequencing data. Nucleic acids research. 2014;42(3):1414-26.

[243] Head SR, Komori HK, LaMere SA, Whisenant T, Van Nieuwerburgh F, Salomon DR, et al. Library construction for next-generation sequencing: overviews and challenges. BioTechniques. 2014;56(2):61-4, 6, 8, passim.

[244] Barrett SP, Salzman J. Circular RNAs: analysis, expression and potential functions. Development (Cambridge, England). 2016;143(11):1838-47.

[245] Szabo L, Salzman J. Detecting circular RNAs: bioinformatic and experimental challenges. Nature reviews Genetics. 2016;17(11):679-92.

[246] Fan X, Zhang X, Wu X, Guo H, Hu Y, Tang F, et al. Single-cell RNA-seq transcriptome analysis of linear and circular RNAs in mouse preimplantation embryos. Genome biology. 2015;16(1):148.

[247] Kim JK, Kolodziejczyk AA, Ilicic T, Teichmann SA, Marioni JC. Characterizing noise structure in single-cell RNA-seq distinguishes genuine from technical stochastic allelic expression. Nature communications. 2015;6:8687.

[248] Jia C, Hu Y, Kelly D, Kim J, Li M, Zhang NR. Accounting for technical noise in differential expression analysis of single-cell RNA sequencing data. Nucleic acids research. 2017;45(19): 10978-88.

[249] Byron SA, Van Keuren-Jensen KR, Engelthaler DM, Carpten JD, Craig DW. Translating RNA sequencing into clinical diagnostics: opportunities and challenges. Nature reviews Genetics. 2016;17(5):257-71.

[250] Hardwick SA, Deveson IW, Mercer TR. Reference standards for next-generation sequencing. Nature reviews Genetics. 2017;18(8):473-84.

[251] Munro SA, Lund SP, Pine PS, Binder H, Clevert DA, Conesa A, et al. Assessing technical performance in differential gene expression experiments with external spike-in RNA control ratio mixtures. Nature communications. 2014;5:5125.

[252] Consortium SM-I. A comprehensive assessment of RNA-seq accuracy, reproducibility and information content by the Sequencing Quality Control Consortium. Nat Biotechnol. 2014;32(9):903-14.

[253] Li S, Tighe SW, Nicolet CM, Grove D, Levy S, Farmerie W, et al. Multi-platform assessment of transcriptome profiling using RNA-seq in the ABRF next-generation sequencing study. Nat Biotechnol. 2014;32(9):915-25.

[254] t Hoen PA, Friedländer MR, Almlöf J, Sammeth M, Pulyakhina I, Anvar SY, et al. Reproducibility of high-throughput mRNA and small RNA sequencing across laboratories. Nat Biotechnol. 2013;31(11):1015-22.

Chapter 5

Diagnostic Applications for RNA-Seq Technology and Transcriptome Analyses in Human Diseases Caused by RNA Viruses

Irina Vlasova-St. Louis, Andrew Gorzalski and Mark Pandori

Abstract

Human diseases caused by single-stranded, positive-sense RNA viruses, are among the deadliest of the 21st Century. In particular, there are two notable standouts: human immunodeficiency virus (HIV) and severe acute respiratory syndrome coronavirus 2 (SARS-CoV-2). Detection of these disease-causing viral transcripts, by next-generation RNA sequencing (RNA-Seq), represents the most immediate opportunity for advances in diagnostic, therapeutic, and preventive applicability in infectious diseases (e.g., AIDS and COVID-19). Moreover, RNA-Seq technologies add significant value to public health studies by first, providing real-time surveillance of known viral strains, and second, by the augmentation of epidemiological databases, construction of annotations and classifications of novel sequence variants. This chapter intends to recapitulate the current knowledge of HIV and SARS-CoV-2 transcriptome architecture, pathogenicity, and some features of the host immune response. Additionally, it provides an overview of recent advances in diagnostic sequencing methodologies and discusses the future challenges and prospects on the utilization of RNA-Seq technologies.

Keywords: Next generation RNA Sequencing, RNA-Seq, RNA viruses, human immunodeficiency virus, severe acute respiratory syndrome coronavirus, HIV transcriptome, SARS-CoV-2 transcriptome

1. Introduction

Next-generation RNA sequencing is rapidly replacing cDNA microarrays and quantitative PCR in clinics and in the public health laboratories, due to higher sensitivity and precision, as well as its cost-effective ability to identify novel RNA species. High-throughput sequencing has become widely adopted in infectious diseases. Supporting bioinformatics services have improved substantially, which aids in genotyping causative mutations for antimicrobial resistance, pathogens' virulence factors, and global epidemiological surveillance to monitor molecular dynamics of pathogen diversity [1, 2]. Integration of pathogen genome sequencing into infectious disease surveillance was established by United States Centers for Disease Control and Prevention (CDC) under Advanced Molecular Detection program in 2016 [2]. As described below, RNA sequencing became particularly important to identify newly emerging HIV and SARS-CoV strains. RNA-Seq can be used for many different purposes, such as improve diagnostics, characterization of novel strains, investigation of disease epidemiology, spatiotemporal spread and transmission routes, assessment of evolutionary rates, and the development of countermeasures and public health policies.

The standard practice of viral RNA detection is composed of a two-step procedure - reverse transcription of RNA into cDNA and cDNA amplification. The reverse transcription reaction is prone to recombination errors leading to potential alterations in the amplicon library and inadequate sequence representation of the original sample. Although both the nucleic acid amplification test (NAAT) and real-time reverse transcriptase - polymerase chain reaction (RT-PCR) test are cost-effective, the occurrence of amplicon failures needs to be monitored for substitutions in primer binding sites. In order to circumvent possible mutations in the primer regions, it is recommended multiple primer designs targeting the same genes, when designing RT-PCR assays [3, 4]. The RNA-Seq method is capable to unbiasedly identify variants across thousands of target regions with single-base resolution, in cases where NAAT or qRT-PCR produce false-negative results.

Illumina's technology appears to be at the forefront of viral sequencing, however others such as IonTorrent and Oxford Nanopore technologies are quickly closing the gaps [5]. Illumina deploys sequencing-by-synthesis chemistry, which provides high accuracy and deep coverage of the viral genome. A limitation of Illumina is that short read length (< 400 nt) fragmented sequences should be re-assembled computationally, during which the haplotype information can be lost. Short-read sequencing data requires computational pipelines for trimming of low-quality reads, removal of optical duplicates, detection of reads orientation, and alignment to reconstruct full-length viral sequences. Short read RNA-Seq technologies allow to pull multiple samples per lane to reduce the cost, however, demultiplexing of reads from different samples can be bioinformatically challenging.

The recent shift toward emerging long-read technologies, enabled direct sequencing of individual RNA molecules as opposed to classical indirect approaches [6]. Long-read RNA-seq technologies, such as that provided by Ion Torrent or Oxford Nanopore, allow mapping of novel insertions, deletions, or substitutions in natural variants [7]. Thus, the technology offers advantages over established sequencing technologies, as it has simplified procedures for library preparation and bioinformatics analysis. Although direct RNA sequencing possesses low accuracy, than Illumina's short read method, it enables easier full-length sequencing, rapid viral genome annotation and analysis, which would be particularly useful for understanding changes in transcriptome architectures [8]. Additionally, long-read technologies are more cost-effective, portable, and provide robust reproducibility of the results, when compared to short-read RNA-seq [9]. A heterozygous variants can be detected when filtered by bioinformatic pipelines, which may identify co-infections and estimate the risk of superinfection [10]. Detailed descriptions and of long- and short- read methodologies, as well as analysis workflows, are provided in other chapters of this book.

2. Sequencing of human immunodeficiency virus (HIV)

The human immunodeficiency virus (HIV) is organized in the genus Lentivirus, within the family of Retroviridae, subfamily Orthoretrovirinae. HIV is a human retrovirus with an RNA genome that is composed of copies of a single stranded positive sense RNA [11]. The RNA genome encodes viral enzymes and is packed into nucleocapsid proteins (the viral capsid). After infecting the cell, the viral RNA genome is reverse transcribed into double stranded viral DNA, which is subsequently integrated into cellular host DNA with the aid of viral and host cofactors. The HIV life cycle has several life-cycle phases; a prolonged latent phase of host genome integration, and a phase of active transcription of new viral RNA in the infected cell [11]. The of HIV replication process is characterized by high rates of nucleotide misincorporations, insertions, deletions, and recombinations. Twentieth century research studies, in 1980s and 1990s, uncovered only a partial viral genome. Determination of the HIV partial genome sequence provided the basis for the development of next-generation sequencing and real-time reverse transcriptase polymerase chain reaction (RT-PCR) methods.

Next-generation RNA-Sequencing enabled the first successful pan-HIV-1 sequencing, including subtype identification and phylodynamic diversification during the course of AIDS pandemic [12]. The current HIV nomenclature includes two types: HIV-1 and HIV-2. The HIV-1 is phylogenetically classified into of 4 groups (M, N, O, and P) [13]. Phylodynamic analyses and evolutionary clock models estimated the time of the most recent common ancestor (HIV-1 pandemic group) to be around the late 19th – early 20th centuries, in Central Africa, and where it may have been associated with an epidemic in primates [14]. The major (M) pandemic group, which causes human-to-human transmission, is classified into 9 subtypes. The most prevalent M subtypes are A, B, C, D, and G; while subtypes F, H, J, L and K are collectively responsible for approximately 1% of all HIV infections [14–16]. An important genetic feature of HIV is that it is prone to recombination. The highest levels of inter-subtype re-combinations are found in HIV-1 infected specimens, from patients in the Sub-Saharan region of Africa. Analysis of a whole-genome HIV-1 sequence, from the Congo Basin, identified ancestral HIV species which clustered basal to all major subtypes; many of which underwent purifying selection and are no longer in circulation. However, 72 additional recombinant forms of HIV remain in circulation [17].

Intra-individual mixtures of recombinant genomes have been reported throughout the world. It is hypothesized that inter-subtype recombinant viruses have an advantage of transmissibility over parental strains [18–24]. To achieve full viral genome sequence coverage, several approaches were employed. One such approach was the amplification of two large fragments ("half genomes"), spanning the full HIV-1 sequence, including all critical regions. Using this approach, molecular surveillance studies of circulating and recombinant forms of HIV, has been conducted in Cameron. Researchers used primers which target two overlapping "half genome" sequencing approach: the 5′ region (*gag* and *pol* genes), the 3′ region (*env* and *nef*), which overlap the accessory genes (*vif, vpr,* and *vpu*) [25]. The two-amplicon approach followed by deep sequencing allowed to characterize several novel circulating recombinant forms, CRFs, (CRF02_AG, _AE, _01A1, and F2, CRF_36cpx and _37cpx, etc.) and more than a dozen unique recombinant forms (URF, NYU6541_6, NYU6556_3, etc.) that clustered separately from their reference sequences, as determined by the maximum likelihood phylogenetic algorithm [26]. The introduction of circulating recombinant forms (CRF01_AE and CRF02_AG) into the Asian-Pacific region from Continental Asia was identified using similar "half genome" sequencing approach [27]. Another approach, the "switching mechanism at the 5′ end of RNA transcript" (SMART), leverages the template-switching capability of certain RT enzymes toward full-length template. During RT reaction, three additional nucleotides are added to the 3′ end of the first cDNA strand of viral RNA template, which serve as an anchor for selective amplification of full length template with a set of 5′ adaptor-ligated primer [28]. This method was adopted to capture full-length HIV sequence in clinical samples, in which viral loads were > 5 log10 copies/mL [29]. Neighbor-joining phylogenetics trees built from this study data revealed all known viral recombinant species from all four groups

(M, N, P, O), and rare M-subtypes (J and H). A successful strategy has been to study HIV species diversity by embedding a degenerate block of nucleotides, within the primer, during first round of cDNA synthesis, to avoid PCR-related errors (and sequencing errors [30, 31]. A unique ID (Primer ID), created by an incorporated primer tag, can correct allelic skewing during sequencing [32]. Using this approach the authors were able to identify minor variants in the HIV-1 protease gene resulting in multiple alleles that conferred drug-resistance [30].

The percentage of people living with HIV drug resistance appears to be increasing [33]. RNA-seq has been key to revealing so-called viral invasive genes and the acquisition of drug-resistance mutations throughout of the 9.7 Kb of viral genome [34]. Achievement of long amplicon length allowed quantification of viral load and multiple drug-resistant mutations using viral RNA and pro-viral DNA as a template [35–38]. RNA-based long-read sequencing is also being explored for surveillance of viral diversification and assessment of HIV quasispecies [39, 40]. The FDA has cleared Sentosa HIV-1 RNA-Seq genotyping SQ assay for clinical use, to detect and monitor antiretroviral drug resistance mutations in the blood of HIV-infected patients. This product helps clinicians to prescribe effective combination of antiretroviral drugs, such as reverse transcriptase, integrase, envelope (gp41), or CCR5/CXCR4 inhibitors [39]. Phylogenetic clustering of genotyped samples from HIV patients who diagnosed between 1999 and 2012 identified strong associations between molecular clusters and epidemiological hotspots for transmitted drug resistance [41].

3. Sequencing of SARS-CoV-2

Viruses of the order Nidovirales - family Coronaviridae - subfamily Coronavirinae are positive-sense, single-stranded sequences of RNAs; ranging from 26 to 32 kilobases in length [42]. The novel SARS-CoV-2 species was identified via metagenomic RNA sequencing and made publicly available, in NCBI Virus, within two months of unknown pneumonia cases outbreak in China (accession number NC_045512) [43]. The SARS-CoV-2 viral genome (made up of 29,903 nucleotides) is phylogenetically related to genus beta-coronavirus (subgenus sarbecovirus), which has demonstrated the ability to crossover, from the animal kingdom, and subsequently cause infection in humans [44]. A viral RNA genome encodes several structural proteins (spike, S; nucleocapsid, N; transmembrane, M; envelope, E). SARS-CoV-2 is known to produce 16 non-structural proteins, and at least six accessory proteins (encoded primarily by reading frames ORF1 and ORF2) [45]. Each viral transcript has a 5′ cap structure and a 3′ poly(A) tail [46]. Genomic characterization of SARS-CoV-2 receptor-binding domains, in conjunction with phylogenetic analysis and homology modeling, revealed the sequence for viral port of entry proteins that conferred human-to-human transmission [47]. Direct RNA sequencing via nanopore and DNA nanoball methodologies identified that SARS-CoV-2 generates a tiered series of canonical and non-canonical subgenomic RNAs, through a process involving homology recombinations between transcriptional regulatory sequences [48]. The function of these transcripts is currently unknown [49].

Sequencing-based genomic surveillance has been employed by UK, USA, and Canadian consortia, to coordinate efforts to sequence large numbers of SARS-CoV-2 genomes. An executive summary "Genomic sequencing of SARS-CoV-2" was issued by the WHO, in January 2021. It discusses the implementation of NGS, for "maximum impact on public health" and provide detailed overview of current sequencing methodologies [50].

In lockstep with the continuing advances of next-generation sequencing technologies, the bioinformatics community has developed many computational tools capable of keeping up with big data analysis and interpretation. Data sharing repositories such as GISAID (EpiCoV), NCBI Virus/GenBank, COG-UK, ENA provide free access to sequencing data (**Figure 1**). Bioinformatics workspaces (e.g., Terra [51], UshER [52], NextStrain [53, 54], PANGO Lineages [55], CoV-Glue [56], see **Figure 1**) allow users to assign their own sequences to globally circulating lineages [57].

Genomics researchers who track phylogenetic dynamics of SARS-CoV-2 developed several schemes to describe the accumulation of mutations and detectable divisions within circulating SARS-CoV-2 populations (**Figure 2**). Three popular clade naming conventions are used:

1. PANGO Lineages [55];

2. Clades by NextStrain [58];

3. Clades by GISAID [59].

The PANGO Lineages nomenclature refers to corresponding outbreaks caused by distinct lineages (**Figure 2A**). Lineage A is suggested to be ancestral because it shares more nucleotides similarities with related coronaviruses in animals [60]. Lineages B presumably originated from United Kingdom, and lineages P – from Brazil [61, 62]. The Clades by NextStrain and GISAID nomenclature refers to the divergence of newly emergent clades (see **Figure 2B** and **C**) [57, 59, 63]. A formal naming system for SARS-CoV-2 evolutionary lineages has not been universally adopted. The nomenclatures from PANGO Lineages and GISAID are updated on CDC and WHO web [64, 65]. Novel phylodynamic models for visualization of phylogenomic datasets have been employed in different web formats. For example, Auspice, Augur, NextClade are open-source interactive web tools for visualizing phylogenomic data within NextStrain interface [66]. UShER (Ultrafast Sample placement on Existing tRee) is available in UCSC, and the virus phylogeny is a part of T-BioInfo platform [52]. Novel human coronavirus lineages are now regularly reported, and lineage nomenclature is continually updated [64]. The WHO Virus Evolution Working Group has been working on simplifying the nomenclature, for SARS-CoV-2 variants of interests (VOI) and variants of

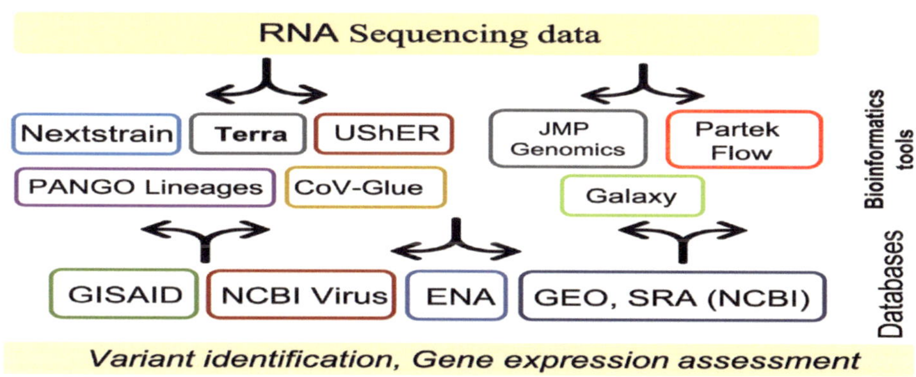

Figure 1.
Diagnostic applications for RNA-Seq data analyses in human diseases caused by RNA viruses. The upper boxes represent bioinformatics workspaces for data analysis as described in the text below. The lower boxes represent major public repositories/databases commonly used for collecting and sharing genome sequence data (see abbreviations list). Note: Viral sequencing data are deposited in GISAID and NCBI virus; the ENA (EMBL-EBI), GEO, and SRA (NCBI) contain sequences from all domains of life, including human specimens). Double-headed arrows represent interactive nature between bioinformatics tools and data repositories, which allows for variants identification and gene expression assessment.

2A. PANGO Lineages.

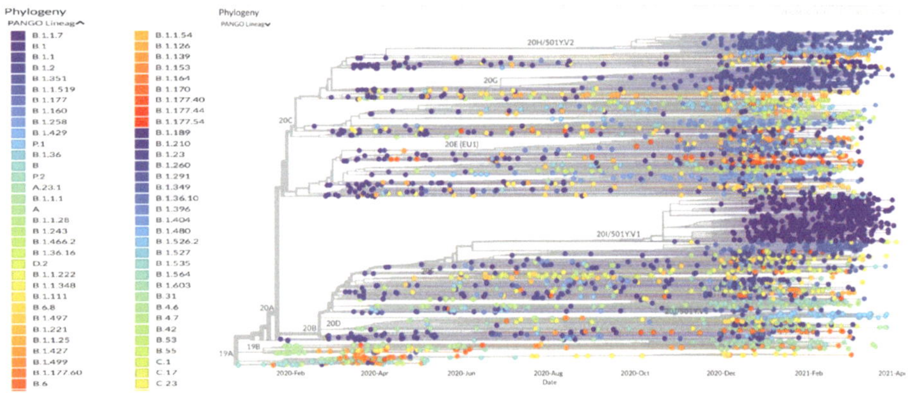

2B. Clades by NextStrain.

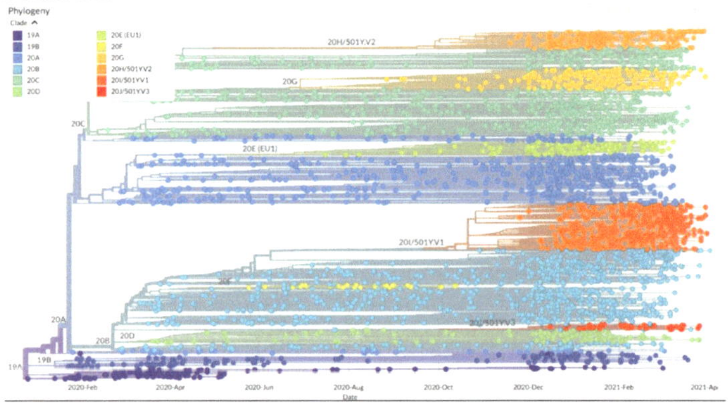

2C. Clades by GISAID.

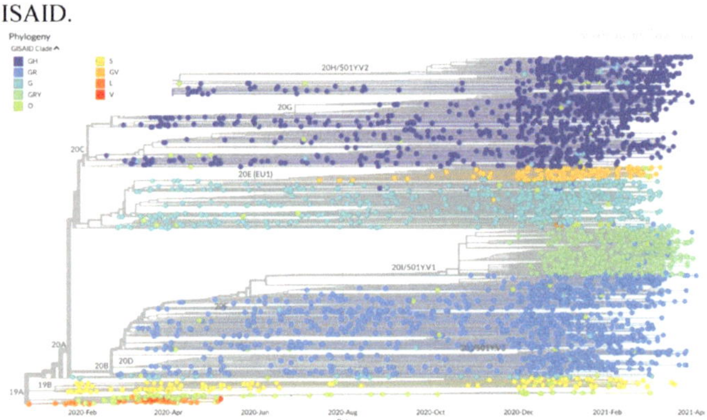

Figure 2.
Genomic epidemiology of novel severe acute respiratory syndrome coronavirus 2 (global subsampling). 2A. PANGO lineages. 2B. Clades by NextStrain. 2C. Clades by GISAID. These phylogenetic trees show time-resolved visualization of evolutionary relationships between SARS-CoV-2 viruses from the ongoing COVID-19 pandemic. Exported from https://nextstrain.org/ncov/global. The dots on the trees' branches are color-coded in accordance with phylogeny classification by (2A) PANGO lineages; (2B) by NextStrain clades; (2C) clades by GISAID. Figure legends on the left represent lineages or clades names. Accessed on Apr 20th.

concern (VOC). This group has recommended easy-to-pronounce variant naming based on the Greek alphabet (e.g. Alpha, Beta, Gamma, Delta - for VOC; and Epsilon thought Lambda – for VOI) [67].

Genomic epidemiology is a novel discipline that assesses viral genomic diversification, including the acquisition of pathogenic mutations, and molecular dynamics of a pandemic, at the population level. Real-time viral sequencing helps delineate the origin of imported cases. For example, the assessment of WGS data (in several European countries) enabled a more precise understanding of SARS-CoV-2 transmission patterns, during various phases of the current pandemic. Molecular epidemiology data aided in identifying multiple introduction events, from the community spread in the Netherlands, Brazil, and several other countries [68–71]. WGS analysis leveraged with phylogenetic methodology approximated the source of lineage exportation. For example, the first outbreak of COVID-19 in Taiwan was imported from China. However, the second was from Italy, as strains that caused the second outbreak were phylogenetically more related to GISAID Accession IDs from the Italian outbreak [72]. As determined by the neighbor-joining method using MEGAX software, the circulating Moroccan lineage is more closely related to the South African lineage (20H/501Y.V2). This is not surprising given that travel within the continent is more predominant than travel between continents [73, 74]. In the United States, epidemiological investigation of sequencing data showed that SARS-CoV-2 varies at the single-nucleotide polymorphism (SNP) resolution, within various States. Additionally, SARS-CoV-2 genomic surveillance identified introduction and community spread of more transmissible strains in the states of California, Ohio, Washington and others (e.g., 20C > 20G clade switch) [75–79]. Genome sequencing in the New York City and Boston areas identified a cryptic spread of multiple simultaneous lineage introductions, from European or Asian travel-related exposures [80–82]. Using Global Epidemic and Mobility Model (GLEAM), Davis et al. estimated the time for the arrival and cryptic phase of the COVID-19 epidemic, which began in early to mid-January on both the East and West coasts of the USA [83]. During the cryptic phase of transmission, monophyletic clades of imported SARS-CoV-2 were spreading until air travel restrictions were dictated [83, 84].

US Center for Disease Controls and Prevention regularly updates the SARS-CoV-2 variant classification, adding novel mutations that have an impact on the immune response or virus virulence factor. Variants are generally classified as a variant of interest, a variant of concern, and a variant of high consequence [64]. Genome-wide sequencing analysis of samples, from various geographic locations, has revealed novel mutations dispersed throughout SARS-CoV-2 genome [85]. Weber et.al has recently detected a number of hotspot mutations, juxtaposing SARS-CoV-2 sequences from various geographic regions, which occurred *de novo*, during the dissemination of infection [60]. These differences in genome variation are not necessarily all pathogenic, but may be important in vaccine design and assessment of immunogenicity [86]. Molecular clock computations estimate that the average evolutionary rate for coronaviruses is roughly 10^{-4} nucleotide substitutions per site, per year, with point mutations that contribute to the diversification of global strains [87, 88]. Although most acquired genetic changes do not substantially affect virulence or transmissibility, those that do cause the corresponding phenotypic changes in SARS-CoV-2 - are of public health importance [67, 85].

4. Advances in diagnostic sequencing methods

Diagnostic sequencing methods have reached a state of capability, which enables the detection of low quantities of viral RNA, with greater sensitivity and specificity. Several amplicon construction approaches have been utilized: targeted capture-based, tiling hexamer-based, and standard amplicon-based [89–91]. A newly commercialized methodology that includes the hybridization capture method yielded near-complete genome coverage, even in samples with relatively low viral loads (cycle thresholds, Ct > 25) [92]. Several studies reported that 93 to 99% of genome coverage is the samples with Ct values ranging from 26 to 32. The median on-target percentage of genome coverage dropped below 50%, in higher Ct value samples (Ct > 30) [93]. Public Health Laboratories have recently utilized capture-based methods followed by sequencing to evaluate if SARS-CoV-2 is present in wastewater samples. The goal of the study was the detection the meta-transcriptomic differences between SARS-CoV-2 variants, found in wastewater, and compare the differences to locally reported clinical genotypes [77]. An Illumina Flex for Enrichment kit and Illumina Respiratory Virus Oligo Panel were used to enrich the samples for virus cDNA amplicon.

The tiling amplicon-based approach has been adapted for SARS-CoV-2 sequencing. It is based on modified reverse transcription (RT with designed primer pool) and an overlapping long amplicon multiplex PCR strategy [94]. Primers can be designed in a random-hexamer-primed RT fashion, and a not-so-random-hexamer-primed fashion, to eliminate human rRNA during cDNA synthesis. The number of primers can reach up to 400, in metagenomics protocols, to improve genome coverage. However, less than 20 primers are used more widely for amplicon construction [95–97]. The tiling amplicon-based approach has been validated by many laboratories and adapted for nanopore sequencing technologies as well as Illumina and Sanger sequencing [97–99]. A Swift Biosciences' Normalase® tiling amplicon SARS-CoV-2 panel achieved 80% success of partial genome recovery from samples with a Ct value between 32 and 34, and 40% - for samples with a Ct value between 34 and 36 [100]. A 98 non-overlapping primer pair set have been designed by a group of scientists from ARTIC network in early March of 2020. This primer set has been modified on 3 occasions, reducing primer-dimers formation during RT-PCR (a major cause of coverage bias) [101]. Recently, the ARTIC's CoronaHiT platform (Coronavirus High Throughput) has been launched and provided accurate consensus SARS-CoV-2 genomes when at least 20x coverage is obtained. It is important to note that in the low-titer samples, with Ct > 32 (< 100 viral genome copies/uL), sequencing results became less reliable for all sequencing methods (long- and short- read) [96, 102–104].

5. Assessment of host response by RNA-Seq

The host-pathogen interactions have been studied by RNA-Seq in research settings. Transcriptome analysis of hosts' diseased tissues has been an integral part of the discovery of immune response biomarkers. University of Minnesota Division of Infectious Diseases (USA) leads intensive transcriptomic studies for discoveries of novel biomarkers of immune reconstitution inflammatory syndrome (IRIS) in the HIV-infected population [105, 106]. IRIS presents as an exaggerated immune reaction (in a form of cytokine release syndrome) that occurs in immunocompromised patients who have commenced antiretroviral treatments (ART) [106]. Several predictive and diagnostic biomarkers have already been identified in peripheral blood, utilizing Galaxy, JMP Genomics, or Partek Flow software (**Figure 1**). For example, the low expression of type I interferon, interferon-response genes, and components of antiviral defense pathways were identified as a risk factor for subsequent occurrence of non-fatal forms of IRIS [107, 108]. More recently, transcriptomic predictors of fatal IRIS have been delineated. These include the overexpressed genes that encode numerous proinflammatory cytokines (e.g., IL6), chemokines, stress response kinase signaling and production of reactive oxygen species in monocytes pathways [109]. Interestingly, the deficiency in innate antiviral defense genes, and poly-immuno-cytopenia have also been identified as significant contributors to subsequent cytokine release syndrome during COVID-19 [110–112]. The reduced IFN I and IFN III type gene expression and elevated monocyte- and granulocyte- derived chemokines at an early stage of

COVID-19 have been proposed as predictors of subsequent cytokine release syndrome and disease severity [113].

In a small percentage of the human population, the SARS-CoV-2 infection results in a critical illness that begins with uncontrolled overexpression of proinflammatory cytokines (the so-called "cytokine storm") [114]. A cytokine storm may lead to pathophysiological events such as activation of the coagulation cascades, systemic oxidative stress, and various forms of cell death [115]. Clinically, these events manifest as an acute respiratory distress syndrome (ARDS), or death from multiple organ failure [116, 117]. SARS-CoV-2 is recognized by innate immune cells and, in cases of ARDS, elicits highly abundant transcriptional gene expression, which is somewhat comparable to that of immunocompromised patients who have developed fatal IRIS [109, 118, 119]. The upregulation of inflammasome, Toll-like receptor signaling, HMGB1, and oxidative stress response transcripts, have been recently demonstrated as biomarkers of fatal immune reconstitution inflammatory syndrome [109]. Additionally, NF-kB-associated inflammatory genes and strong neutrophile activation/degranulation signatures differentiate fatal IRIS events from those of survivors. Likewise, fatal ARDS in COVID-19 patients have been characterized by maladapted hyperactivation of the same innate inflammatory pathways described for fatal IRIS [120–123]. Systemic upregulation of cytokines TNFa, IL6, IL1B, and NLRP3 are noteworthy of mention as they are targetable molecules for drugs such as adalimumab, tocilizumab, anakinra, and RAPA-501-Allo, respectively [124–127].

SARS-CoV-2 has the propensity to selectively induce mortality in elderly population (over 65 years of age) and in immunocompromised individuals [128, 129]. The most reasonable explanation for this propensity is the depletion of the thymic population, and the inability to mount adaptive T cell immune responses in timely manner [130]. Innate immune cells become the first line of defense against SARS-CoV-2, however in the absence of proper regulatory feedback – are thrown into an uncontrolled proinflammatory loop. Similarly, the irreparable thymic damage and a waning adaptive immunity in AIDS patients enable the sustained HIV replication, which may explain why serious complications such as fatal IRIS are hinged on responses driven by innate immune cells.

There have been isolated cases of SARS-CoV-2 reinfection in individuals who had no known immune disorders [131–133]. Since SARS-CoV-2 has not diversified significantly, these patients have been re-infected with variants from the same or adjacent clade [132]. Studies have revealed a number of SARS-CoV-2 epitopes are recognized by CD8+ T cells in COVID-19 convalescent subjects (as determined by TruSight HLA sequencing panel) [134]. It is hypothesized that re-infected individuals are unable to mount a sufficient response to primary infection, which result in secondary COVID-19 manifestation. Many of these patients had tested positive for anti- SARS-CoV-2 antibodies after the reinfection, but it is unknown whether they developed antibodies after the first infection at all. It has been hypothesized that susceptibility to re-infection is genetically driven [135]. Several hypothetical models had been proposed [136]. Thus, simultaneous RNA-seq of host and pathogen during viral infection would be helpful to characterize the molecular responses in case of COVID-19 reinfection.

There are now several vaccine types that are being used, in an attempt to reach herd immunity, in various countries [137]. The most successful vaccines are mRNA-based, which have an estimated 92% efficacy [138, 139]. Still, there is an accumulating body of evidence of the so-called 'vaccine breakthrough' phenomenon, when SARS-CoV-2 infection occurs in persons who are fully vaccinated [140, 141]. The immune gene variants that contribute to poor T cell memory and undelaying vaccine breakthroughs are not yet known [142]. As vaccination increases, RNA-Seq may become a useful tool in exploring the consequence of vaccine dose-spearing strategies, the emergence of novel vaccine-escape variants, intra- host variant diversity, as well as in understanding the phenomenon of vaccine breakthrough.

6. Conclusion and future directions

RNA-seq has become an indispensable diagnostic tool of clinically important RNA viruses. It has taken the research community almost 20 years to sequence a complete HIV viral genome, but it took a little over 2 months to sequence a near complete SARS-CoV-2 genome and identify its origin! With an improved understanding of the genomic structures of virus populations, RNA-seq can be used to track the origins, genetic diversity, and global monitoring of transmission patterns. In a public health context, RNA-Seq is an indispensable tool to be utilized in the assessment of risks and the emergence of future outbreaks. Additionally, assessment of transcriptomic profiles, during host responses to infection improved the understanding of disease pathogenesis and identify patients at risk for severe outcomes. To overcome drug resistance, novel viral genotype-guided agents based on antisense RNAs (e.g., aptamers, peptide nucleic acids oligomers, ribozymes, and RNAi silencing therapies) are being developed to specifically inhibit viral replication. Complex bioinformatics approaches enabled vaccines to be designed with high probabilities of intercepting viral escape [143]. Much work awaits the global scientific community, including the establishment of unified reference standards and the adoption of a single nomenclature for SARS-CoV-2. Additionally, the assessment of long-term vaccination efficacy by RNA-Seq in the clinical laboratory is necessary to fully understand its benefits [144].

Abbreviations

SARS-CoV-2	severe acute respiratory syndrome coronavirus 2
CDC	Center for Disease Controls and Prevention
UCSC	University of California, Santa Cruz
GISAID	Global Initiative on Sharing Avian Flu Data
ENA	European Nucleotide Archive
COG-UK	COVID-19 Genomics UK Consortium
WGS	whole genome sequencing
SNP	single-nucleotide polymorphism
NCBI	The National Center for Biotechnology Information
UShER	Ultrafast Sample placement on Existing tRee
IRIS	immune reconstitution inflammatory syndrome
TNFA	Tumor Necrosis Factor Alfa
HMGB1	High Mobility Group Box 1 gene
NLRP3	NLR (nucleotide-binding domain and leucine-rich repeat containing) Family Pyrin Domain Containing 3
EMBL-EBI	European Molecular Biology Laboratory - European Bioinformatics Institute

Author details

Irina Vlasova-St. Louis[1,2]*, Andrew Gorzalski[2] and Mark Pandori[2]

1 Department of Medicine, University of Minnesota, Minneapolis, MN, USA

2 Nevada State Public Health Laboratory, Reno, NV, USA

*Address all correspondence to: irinastl@umn.edu and istlouis@med.unr.edus

References

[1] Avery, M.; Mills, S.J.; Stephan, E. Real-time monitoring through the use of technology to enhance performances throughout HIV cascades. *Curr. Opin. HIV AIDS* 2017, *12*. 10.1097/COH.0000000000000397.

[2] Canters for Disease Control and Prevention Advanced molecular detection (AMD) Available online: https://www.cdc.gov/amd/index.html (accessed on Mar 16, 2021).

[3] Vogels, C.B.F.; Brito, A.F.; Wyllie, A.L.; Fauver, J.R.; Ott, I.M.; Kalinich, C.C.; Petrone, M.E.; Casanovas-Massana, A.; Catherine Muenker, M.; Moore, A.J.; et al. Analytical sensitivity and efficiency comparisons of SARS-CoV-2 RT–qPCR primer–probe sets. *Nat. Microbiol.* **2020**, *5*. 10.1038/s41564-020-0761-6.

[4] Bustin, S.A.; Nolan, T. RT-QPCR testing of SARS-COV-2: A primer. *Int. J. Mol. Sci.* **2020**, *21*. 10.3390/ijms21083004.

[5] Bull, R.A.; Adikari, T.N.; Ferguson, J.M.; Hammond, J.M.; Stevanovski, I.; Beukers, A.G.; Naing, Z.; Yeang, M.; Verich, A.; Gamaarachchi, H.; et al. Analytical validity of nanopore sequencing for rapid SARS-CoV-2 genome analysis. *Nat. Commun.* **2020**, *11*. 10.1038/s41467-020-20075-6.

[6] Viehweger, A.; Krautwurst, S.; Lamkiewicz, K.; Madhugiri, R.; Ziebuhr, J.; Hölzer, M.; Marz, M. Direct RNA nanopore sequencing of full-length coronavirus genomes provides novel insights into structural variants and enables modification analysis. *Genome Res.* **2019**, *29*. 10.1101/gr.247064.118.

[7] Lee, E.R.; Parkin, N.; Jennings, C.; Brumme, C.J.; Enns, E.; Casadellà, M.; Howison, M.; Coetzer, M.; Avila-Rios, S.; Capina, R.; et al. Performance comparison of next generation sequencing analysis pipelines for HIV-1 drug resistance testing. *Sci. Rep.* **2020**, *10*. 10.1038/s41598-020-58544-z.

[8] Gwinn, M.; Maccannell, D.; Armstrong, G.L. Next-Generation Sequencing of Infectious Pathogens. *JAMA - J. Am. Med. Assoc.* 2019, *321*. 10.1001/jama.2018.21669.

[9] Helmy, Y.A.; Fawzy, M.; Elaswad, A.; Sobieh, A.; Kenney, S.P.; Shehata, A.A. The COVID-19 Pandemic: A Comprehensive Review of Taxonomy, Genetics, Epidemiology, Diagnosis, Treatment, and Control. *J. Clin. Med.* **2020**, *9*. 10.3390/jcm9041225.

[10] Sjaarda, C.P.; Rustom, N.; Evans, G.A.; Huang, D.; Perez-Patrigeon, S.; Hudson, M.L.; Wong, H.; Sun, Z.; Guan, T.H.; Ayub, M.; et al. Phylogenomics reveals viral sources, transmission, and potential superinfection in early-stage COVID-19 patients in Ontario, Canada. *Sci. Rep.* **2021**, *11*. 10.1038/s41598-021-83355-1.

[11] Seitz, R. Human Immunodeficiency Virus (HIV). *Transfus. Med. Hemotherapy* **2004**, *31*. 10.1159/000078043.

[12] Etienne, L.; Delaporte, E.; Peeters, M. Origin and Emergence of HIV/AIDS. In *Genetics and Evolution of Infectious Diseases*;2011.10.1016/B978-0-12-384890-1.00026-1.

[13] Castro-Nallar, E.; Pérez-Losada, M.; Burton, G.F.; Crandall, K.A. The evolution of HIV: Inferences using phylogenetics. *Mol. Phylogenet. Evol.* 2012, *62*. 10.1016/j.ympev.2011.11.019.

[14] Gryseels, S.; Watts, T.D.; Mpolesha, J.M.K.; Larsen, B.B.; Lemey, P.; Muyembe-Tamfum, J.J.; Teuwen, D.E.; Worobey, M. A near full-length HIV-1 genome from 1966 recovered from formalin-fixed paraffin-embedded tissue. *Proc. Natl. Acad. Sci. U. S. A.* **2020**, *117*. 10.1073/pnas.1913682117.

[15] LANL HIV database Available online: https://www.hiv.lanl.gov/content/index (accessed on Mar 3, 2021).

[16] Souza, J.S.M.; Silva Júnior, J.J.; Brites, C.; Monteiro-Cunha, J.P. Molecular and geographic characterization of hiv-1 bf recombinant viruses. *Virus Res.* **2019**, *270*. 10.1016/j.virusres.2019.197650.

[17] Rodgers, M.A.; Wilkinson, E.; Vallari, A.; McArthur, C.; Sthreshley, L.; Brennan, C.A.; Cloherty, G.; de Oliveira, T. Sensitive Next-Generation Sequencing Method Reveals Deep Genetic Diversity of HIV-1 in the Democratic Republic of the Congo. *J. Virol.* **2017**, *91*. 10.1128/jvi.01841-16.

[18] Sierra, M.; Thomson, M.M.; Ríos, M.; Casado, G.; Ojea-de Castro, R.; Delgado, E.; Echevarría, G.; Muñoz, M.; Colomina, J.; Carmona, R.; et al. The analysis of near full-length genome sequences of human immunodeficiency virus type 1 BF intersubtype recombinant viruses from Chile, Venezuela and Spain reveals their relationship to diverse lineages of recombinant viruses related to CRF12_BF. *Infect. Genet. Evol.* **2005**, *5*. 10.1016/j.meegid.2004.07.010.

[19] Bello, G.; Aulicino, P.C.; Ruchansky, D.; Guimarães, M.L.; Lopez-Galindez, C.; Casado, C.; Chiparelli, H.; Rocco, C.; Mangano, A.; Sen, L.; et al. Phylodynamics of HIV-1 Circulating Recombinant Forms 12_BF and 38_BF in Argentina and Uruguay. *Retrovirology* **2010**, *7*. 10.1186/1742-4690-7-22.

[20] Ruchansky, D.; Casado, C.; Russi, J.C.; Arbiza, J.R.; Lopez-Galindez, C. Identification of a new HIV Type 1 circulating recombinant form (CRF38-BF1) in Uruguay. *AIDS Res. Hum. Retroviruses* **2009**, *25*. 10.1089/aid.2008.0248.

[21] Shcherbakova, N.S.; Shalamova, L.A.; Delgado, E.; Fernández-García, A.; Vega, Y.; Karpenko, L.I.; Ilyichev, A.A.; Sokolov, Y. V.; Shcherbakov, D.N.; Pérez-Álvarez, L.; et al. Short communication: Molecular epidemiology, phylogeny, and phylodynamics of CRF63-02A1, a recently originated HIV-1 circulating recombinant form spreading in Siberia. *AIDS Res. Hum. Retroviruses* **2014**, *30*. 10.1089/aid.2014.0075.

[22] Takebe, Y.; Liao, H.; Hase, S.; Uenishi, R.; Li, Y.; Li, X.J.; Han, X.; Shang, H.; Kamarulzaman, A.; Yamamoto, N.; et al. Reconstructing the epidemic history of HIV-1 circulating recombinant forms CRF07_BC and CRF08_BC in East Asia: The relevance of genetic diversity and phylodynamics for vaccine strategies. *Vaccine* **2010**, *28*. 10.1016/j.vaccine.2009.07.101.

[23] Pessôa, R.; Loureiro, P.; Lopes, M.E.; Carneiro-Proietti, A.B.F.; Sabino, E.C.; Busch, M.P.; Sanabani, S.S. Ultra-deep sequencing of HIV-1 near full-length and partial proviral genomes reveals high genetic diversity among Brazilian blood donors. *PLoS One* **2016**, *11*. 10.1371/journal.pone.0152499.

[24] Cañada, J.E.; Delgado, E.; Gil, H.; Sánchez, M.; Benito, S.; García-Bodas, E.; Gómez-González, C.; Canut-Blasco, A.; Portu-Zapirain, J.; Sáez de Adana, E.; et al. Identification of a New HIV-1 BC Intersubtype Circulating Recombinant Form (CRF108_BC) in Spain. *Viruses* **2021**, *13*. 10.3390/v13010093.

[25] Banin, A.N.; Tuen, M.; Bimela, J.S.; Tongo, M.; Zappile, P.; Khodadadi-Jamayran, A.; Nanfack, A.J.; Meli, J.; Wang, X.; Mbanya, D.; et al. Development of a versatile, near full genome amplification and sequencing approach for a broad variety of HIV-1 group M variants. *Viruses* **2019**, *11*. 10.3390/v11040317.

[26] Banin, A.N.; Tuen, M.; Bimela, J.S.; Tongo, M.; Zappile, P.; Khodadadi-Jamayran, A.; Nanfack, A.J.; Okonko, I.O.; Meli, J.; Wang, X.; et al. Near full genome characterization of HIV-1 unique recombinant forms in Cameroon reveals dominant CRF02_AG and F2 recombination patterns. **2019**. 10.1002/jia2.25362/full.

[27] Chen, Y.; Hora, B.; DeMarco, T.; Berba, R.; Register, H.; Hood, S.; Carter, M.; Stone, M.; Pappas, A.; Sanchez, A.M.; et al. Increased predominance of HIV-1 CRF01_AE and its recombinants in the Philippines. *J. Gen. Virol.* **2019**, *100*. 10.1099/jgv.0.001198.

[28] Zhu, Y.Y.; Machleder, E.M.; Chenchik, A.; Li, R.; Siebert, P.D. Reverse transcriptase template switching: A SMART™ approach for full-length cDNA library construction. *Biotechniques* 2001, *30*. 10.2144/01304pf02.

[29] Berg, M.G.; Yamaguchi, J.; Alessandri-Gradt, E.; Tell, R.W.; Plantier, J.C.; Brennan, C.A. A pan-HIV strategy for complete genome sequencing. *J. Clin. Microbiol.* **2016**, *54*. 10.1128/JCM.02479-15.

[30] Jabara, C.B.; Jones, C.D.; Roach, J.; Anderson, J.A.; Swanstrom, R. Accurate sampling and deep sequencing of the HIV-1 protease gene using a Primer ID. *Proc. Natl. Acad. Sci. U. S. A.* **2011**, *108*. 10.1073/pnas.1110064108.

[31] Boltz, V.F.; Rausch, J.; Shao, W.; Coomer, C.; Mellors, J.W.; Kearney, M.F.; Coffin, J.M. Analysis of Resistance Haplotypes Using Primer IDs and Next Gen Sequencing of HIV RNA Methods Available online: http://www.croiconference.org/sites/default/files/posters-2015/593.pdf (accessed on Jul 19, 2015).

[32] Dennis, A.M.; Zhou, S.; Sellers, C.J.; Learner, E.; Potempa, M.; Cohen, M.S.; Miller, W.C.; Eron, J.J.; Swanstrom, R. Using primer-ID deep sequencing to detect recent human immunodeficiency virus type 1 infection. *J. Infect. Dis.* **2018**, *218*. 10.1093/infdis/jiy426.

[33] Porter, D.P.; Daeumer, M.; Thielen, A.; Chang, S.; Martin, R.; Cohen, C.; Miller, M.D.; White, K.L. Emergent HIV-1 drug resistance mutations were not present at low-frequency at baseline in non-nucleoside reverse transcriptase inhibitor-treated subjects in the STaR study. *Viruses* **2015**, *7*. 10.3390/v7122943.

[34] Ekici, H.; Rao, S.D.; Sönnerborg, A.; Ramprasad, V.L.; Gupta, R.; Neogi, U. Cost-efficient HIV-1 drug resistance surveillance using multiplexed high-throughput amplicon sequencing: Implications for use in low- and middle-income countries. *J. Antimicrob. Chemother.* **2014**, *69*. 10.1093/jac/dku278.

[35] Fogel, J.M.; Bonsall, D.; Cummings, V.; Bowden, R.; Golubchik, T.; De Cesare, M.; Wilson, E.A.; Gamble, T.; Del Rio, C.; Batey, D.S.; et al. Performance of a high-throughput next-generation sequencing method for analysis of HIV drug resistance and viral load. *J. Antimicrob. Chemother.* **2020**, *75*. 10.1093/jac/dkaa352.

[36] Novitsky, V.; Zahralban-Steele, M.; McLane, M.F.; Moyo, S.; Van Widenfelt, E.; Gaseitsiwe, S.; Makhema, J.; Essexa, M. Long-range HIV genotyping using viral RNA and proviral DNA for analysis of HIV drug resistance and HIV clustering. *J. Clin. Microbiol.* **2015**, *53*. 10.1128/JCM.00756-15.

[37] Gall, A.; Ferns, B.; Morris, C.; Watson, S.; Cotten, M.; Robinson, M.; Berry, N.; Pillay, D.; Kellam, P. Universal amplification, next-generation sequencing, and assembly of HIV-1 genomes. *J. Clin. Microbiol.* **2012**, *50*. 10.1128/JCM.01516-12.

[38] Taylor, T.; Lee, E.R.; Nykoluk, M.; Enns, E.; Liang, B.; Capina, R.; Gauthier, M.K.; Domselaar, G. Van; Sandstrom, P.; Brooks, J.; et al. A MiSeq-HyDRA platform for enhanced HIV drug resistance genotyping and surveillance. *Sci. Rep.* **2019**, *9*. 10.1038/s41598-019-45328-3.

[39] Sidhu, G.; Schuster, L.; Liu, L.; Tamashiro, R.; Li, E.; Langaee, T.; Wagner, R.; Wang, G.P. A single variant sequencing method for sensitive and quantitative detection of HIV-1

minority variants. *Sci. Rep.* **2020**, *10*. 10.1038/s41598-020-65085-y.

[40] Kijak, G.H.; Sanders-Buell, E.; Harbolick, E.A.; Pham, P.; Chenine, A.L.; Eller, L.A.; Rono, K.; Robb, M.L.; Michael, N.L.; Kim, J.H.; et al. Targeted deep sequencing of HIV-1 using the IonTorrentPGM platform. *J. Virol. Methods* **2014**, *205*. 10.1016/j.jviromet.2014.04.017.

[41] Chung, Y.S.; Choi, J.Y.; Yoo, M.S.; Seong, J.H.; Choi, B.S.; Kang, C. Phylogenetic transmission clusters among newly diagnosed antiretroviral drug-naïve patients with human immunodeficiency virus-1 in Korea: A study from 1999 to 2012. *PLoS One* **2019**, *14*. 10.1371/journal.pone.0217817.

[42] Pal, M.; Berhanu, G.; Desalegn, C.; Kandi, V. Severe Acute Respiratory Syndrome Coronavirus-2 (SARS-CoV-2): An Update. *Cureus* **2020**. 10.7759/cureus.7423.

[43] NCBI Severe acute respiratory syndrome coronavirus 2 isolate Wuhan-Hu-1, complete genome Available online: https://www.ncbi.nlm.nih.gov/nuccore/1798174254 (accessed on Apr 18, 2021).

[44] Gorbalenya, A.E.; Baker, S.C.; Baric, R.S.; de Groot, R.J.; Drosten, C.; Gulyaeva, A.A.; Haagmans, B.L.; Lauber, C.; Leontovich, A.M.; Neuman, B.W.; et al. The species Severe acute respiratory syndrome-related coronavirus: classifying 2019-nCoV and naming it SARS-CoV-2. *Nat. Microbiol.* 2020, *5*. 10.1038/s41564-020-0695-z.

[45] Michel, C.J.; Mayer, C.; Poch, O.; Thompson, J.D. Characterization of accessory genes in coronavirus genomes. *Virol. J.* **2020**, *17*. 10.1186/s12985-020-01402-1.

[46] de Breyne, S.; Vindry, C.; Guillin, O.; Condé, L.; Mure, F.; Gruffat, H.; Chavatte, L.; Ohlmann, T. Translational control of coronaviruses. *Nucleic Acids Res.* 2020, *48*. 10.1093/nar/gkaa1116.

[47] Lu, R.; Zhao, X.; Li, J.; Niu, P.; Yang, B.; Wu, H.; Wang, W.; Song, H.; Huang, B.; Zhu, N.; et al. Genomic characterisation and epidemiology of 2019 novel coronavirus: implications for virus origins and receptor binding. *Lancet* **2020**, *395*. 10.1016/S0140-6736(20)30251-8.

[48] Kim, D.; Lee, J.Y.; Yang, J.S.; Kim, J.W.; Kim, V.N.; Chang, H. The Architecture of SARS-CoV-2 Transcriptome. *Cell* **2020**, *181*. 10.1016/j.cell.2020.04.011.

[49] Nomburg, J.; Meyerson, M.; DeCaprio, J.A. Pervasive generation of non-canonical subgenomic RNAs by SARS-CoV-2. *Genome Med.* **2020**, *12*. 10.1186/s13073-020-00802-w.

[50] World Health Organization Genomic sequencing of SARS-CoV-2. A guide to implementation for maximum impact on public health Available online: https://apps.who.int/iris/bitstream/handle/10665/338480/9789240018440-eng.pdf?sequence=1&isAllowed=y.

[51] Terra Terra.bio Available online: https://app.terra.bio/ (accessed on May 5, 2021).

[52] University of California Santa Cruze UShER: Ultrafast Sample placement on Existing tRee Available online: https://genome.ucsc.edu/cgi-bin/hgPhyloPlace (accessed on Mar 21, 2021).

[53] Hadfield, J.; Megill, C.; Bell, S.M.; Huddleston, J.; Potter, B.; Callender, C.; Sagulenko, P.; Bedford, T.; Neher, R.A. NextStrain: Real-time tracking of pathogen evolution. *Bioinformatics* **2018**, *34*. 10.1093/bioinformatics/bty407.

[54] Nextstrain Genomic epidemiology of novel coronavirus - Global subsampling Available online: https://nextstrain.org/ncov/global?l=radial.

[55] PANGO PANGO lineages Available online: https://cov-lineages.org/ (accessed on Mar 16, 2021).

[56] Singer, J.; Gifford, R.; Cotten, M.; Robertson, D. CoV-GLUE: A Web Application for Tracking SARS-CoV-2 Genomic Variation. *Preprints* **2020**.

[57] Rambaut, A.; Holmes, E.C.; O'Toole, Á.; Hill, V.; McCrone, J.T.; Ruis, C.; du Plessis, L.; Pybus, O.G. A dynamic nomenclature proposal for SARS-CoV-2 lineages to assist genomic epidemiology. *Nat. Microbiol.* **2020**, *5*. 10.1038/s41564-020-0770-5.

[58] Nextstrain Nextstrain SARS-CoV-2 resources Available online: https://nextstrain.org/sars-cov-2 (accessed on Mar 28, 2021).

[59] GISAID Clade and lineage nomenclature Available online: https://www.gisaid.org/references/statements-clarifications/clade-and-lineage-nomenclature-aids-in-genomic-epidemiology-of-active-hcov-19-viruses/#:~:text=Clade definitions in GISAID are augmented with more detailed lineages,of the pandemic strain causing (accessed on Mar 28, 2021).

[60] Zhou, P.; Yang, X. Lou; Wang, X.G.; Hu, B.; Zhang, L.; Zhang, W.; Si, H.R.; Zhu, Y.; Li, B.; Huang, C.L.; et al. A pneumonia outbreak associated with a new coronavirus of probable bat origin. *Nature* **2020**, *579*, 270-273. 10.1038/s41586-020-2012-7.

[61] Virology.org Preliminary genomic characterisation of an emergent SARS-CoV-2 lineage in the UK defined by a novel set of spike mutations.

[62] virology.org Spike E484K mutation in the first SARS-CoV-2 reinfection case confirmed in Brazil, 2020 Available online: https://virological.org/t/spike-e484k-mutation-in-the-first-sars-cov-2-reinfection-case-confirmed-in-brazil-2020/584 (accessed on Mar 29, 2021).

[63] Li, Q.; Guan, X.; Wu, P.; Wang, X.; Zhou, L.; Tong, Y.; Ren, R.; Leung, K.S.M.; Lau, E.H.Y.; Wong, J.Y.; et al. Early Transmission Dynamics in Wuhan, China, of Novel Coronavirus–Infected Pneumonia. *N. Engl. J. Med.* **2020**, *382*. 10.1056/nejmoa2001316.

[64] Centers for Disease Control and Prevention SARS-CoV-2 Variants Available online: https://www.cdc.gov/coronavirus/2019-ncov/cases-updates/variant-surveillance/variant-info.html (accessed on Mar 18, 2021).

[65] World Health Organization WHO | SARS-CoV-2 Variants. *Who* 2020.

[66] Nextstrain Auspice 2.24.1 Available online: https://docs.nextstrain.org/projects/auspice/en/latest/index.html# (accessed on Mar 25, 2021).

[67] WHO Tracking SARS-CoV-2 variants Available online: https://www.who.int/en/activities/tracking-SARS-CoV-2-variants/ (accessed on Jun 17, 2021).

[68] Oude Munnink, B.B.; Nieuwenhuijse, D.F.; Stein, M.; O'Toole, Á.; Haverkate, M.; Mollers, M.; Kamga, S.K.; Schapendonk, C.; Pronk, M.; Lexmond, P.; et al. Rapid SARS-CoV-2 whole-genome sequencing and analysis for informed public health decision-making in the Netherlands. *Nat. Med.* **2020**, *26*. 10.1038/s41591-020-0997-y.

[69] Toovey, O.T.R.; Harvey, K.N.; Bird, P.W.; Tang, J.W.-T.W.-T. Introduction of Brazilian SARS-CoV-2 484K.V2 related variants into the UK. *J. Infect.* **2021**. 10.1016/j.jinf.2021.01.025.

[70] Volz, E.; Hill, V.; McCrone, J.T.; Price, A.; Jorgensen, D.; O'Toole, Á.; Southgate, J.; Johnson, R.; Jackson, B.; Nascimento, F.F.; et al. Evaluating the Effects of SARS-CoV-2 Spike Mutation D614G on Transmissibility and Pathogenicity. *Cell* **2021**, *184*. 10.1016/j.cell.2020.11.020.

[71] Alteri, C.; Cento, V.; Piralla, A.; Costabile, V.; Tallarita, M.; Colagrossi, L.; Renica, S.; Giardina, F.; Novazzi, F.; Gaiarsa, S.; et al. Genomic epidemiology of SARS-CoV-2 reveals multiple lineages and early spread of SARS-CoV-2 infections in Lombardy, Italy. *Nat. Commun.* **2021**, *12*. 10.1038/s41467-020-20688-x.

[72] Liu, J.Y.; Chen, T.J.; Hwang, S.J. Analysis of imported cases of covid-19 in taiwan: A nationwide study. *Int. J. Environ. Res. Public Health* **2020**, *17*. 10.3390/ijerph17093311.

[73] Lemriss, S.; Souiri, A.; Amar, N.; Lemzaoui, N.; Mestoui, O.; Labioui, M.; Ouaariba, N.; Jibjibe, A.; Yartaoui, M.; Chahmi, M.; et al. Complete Genome Sequence of a 2019 Novel Coronavirus (SARS-CoV-2) Strain Causing a COVID-19 Case in Morocco. *Microbiol. Resour. Announc.* **2020**, *9*. 10.1128/mra.00633-20.

[74] Pearson, C.A.; Russell, T.W.; Davies, N.G.; Kucharski, A.J.; CMMID COVID-19 working group; Edmunds, W.J.; Eggo, R.M. Estimates of severity and transmissibility of novel South Africa SARS-CoV-2 variant 501Y.V2. *Preprint* **2021**, *50*.

[75] Zhang, W.; Davis, B.D.; Chen, S.S.; Sincuir Martinez, J.M.; Plummer, J.T.; Vail, E. Emergence of a Novel SARS-CoV-2 Variant in Southern California. *JAMA* **2021**. 10.1001/jama.2021.1612.

[76] Zhang, W.; Govindavari, J.P.; Davis, B.; Chen, S.; Kim, J.T.; Song, J.; Lopategui, J.; Plummer, J.T.; Vail, E. Analysis of SARS-CoV-2 genomes from southern California reveals community transmission pathways in the early stage of the US COVID-19 pandemic. *medRxiv* 2020. 10.1101/2020.06.12.20129999.

[77] Crits-Christoph, A.; Kantor, R.S.; Olm, M.R.; Whitney, O.N.; Al-Shayeb, B.; Lou, Y.C.; Flamholz, A.; Kennedy, L.C.; Greenwald, H.; Hinkle, A.; et al. Genome sequencing of sewage detects regionally prevalent SARS-CoV-2 variants. *MBio* **2021**, *12*. 10.1128/mBio.02703-20.

[78] Bedford, T.; Greninger, A.L.; Roychoudhury, P.; Starita, L.M.; Famulare, M.; Huang, M.L.; Nalla, A.; Pepper, G.; Reinhardt, A.; Xie, H.; et al. Cryptic transmission of SARS-CoV-2 in Washington state. *Science (80-.).* **2020**, *370*. 10.1126/SCIENCE.ABC0523.

[79] Deng, X.; Gu, W.; Federman, S.; du Plessis, L.; Pybus, O.G.; Faria, N.R.; Wang, C.; Yu, G.; Bushnell, B.; Pan, C.Y.; et al. Genomic surveillance reveals multiple introductions of SARS-CoV-2 into Northern California. *Science (80-.).* **2020**, *369*. 10.1126/science.abb9263.

[80] Maurano, M.T.; Ramaswami, S.; Zappile, P.; Dimartino, D.; Boytard, L.; Ribeiro-Dos-Santos, A.M.; Vulpescu, N.A.; Westby, G.; Shen, G.; Feng, X.; et al. Sequencing identifies multiple early introductions of SARS-CoV-2 to the New York City region. *Genome Res.* **2020**, *31*. 10.1101/gr.266676.120.

[81] Gonzalez-Reiche, A.S.; Hernandez, M.M.; Sullivan, M.J.; Ciferri, B.; Alshammary, H.; Obla, A.; Fabre, S.; Kleiner, G.; Polanco, J.; Khan, Z.; et al. Introductions and early spread of SARS-CoV-2 in the New York City area. *Science (80-.).* **2020**, *369*. 10.1126/science.abc1917.

[82] Lemieux, J.E.; Siddle, K.J.; Shaw, B.M.; Loreth, C.; Schaffner, S.F.; Gladden-Young, A.; Adams, G.; Fink, T.; Tomkins-Tinch, C.H.; Krasilnikova, L.A.; et al. Phylogenetic analysis of SARS-CoV-2 in Boston highlights the impact of superspreading events. *Science (80-.).* **2021**, *371*. 10.1126/science.abe3261.

[83] Davis, J.T.; Chinazzi, M.; Perra, N.; Mu, K.; Piontti, A.P. y.; Ajelli, M.; Dean, N.E.; Gioannini, C.; Litvinova, M.; Merler, S.; et al. Estimating the establishment of local transmission and the cryptic phase of the COVID-19

pandemic in the USA. *medRxiv* 2020. 10.1101/2020.07.06.20140285.

[84] Deng, X.; Gu, W.; Federman, S.; du Plessis, L.; Pybus, O.G.; Faria, N.; Wang, C.; Yu, G.; Pan, C.Y.; Guevara, H.; et al. A genomic survey of SARS-CoV-2 reveals multiple introductions into Northern California without a predominant lineage. *medRxiv* 2020. 10.1101/2020.03.27.20044925.

[85] Islam, M.R.; Hoque, M.N.; Rahman, M.S.; Alam, A.S.M.R.U.; Akther, M.; Puspo, J.A.; Akter, S.; Sultana, M.; Crandall, K.A.; Hossain, M.A. Genome-wide analysis of SARS-CoV-2 virus strains circulating worldwide implicates heterogeneity. *Sci. Rep.* **2020**, *10*. 10.1038/s41598-020-70812-6.

[86] Weber, S.; Ramirez, C.; Doerfler, W. Signal hotspot mutations in SARS-CoV-2 genomes evolve as the virus spreads and actively replicates in different parts of the world. *Virus Res.* **2020**, *289*. 10.1016/j.virusres.2020.198170.

[87] van Dorp, L.; Acman, M.; Richard, D.; Shaw, L.P.; Ford, C.E.; Ormond, L.; Owen, C.J.; Pang, J.; Tan, C.C.S.; Boshier, F.A.T.; et al. Emergence of genomic diversity and recurrent mutations in SARS-CoV-2. *Infect. Genet. Evol.* **2020**, *83*. 10.1016/j.meegid.2020.104351.

[88] Liu, Q.; Zhao, S.; Shi, C.M.; Song, S.; Zhu, S.; Su, Y.; Zhao, W.; Li, M.; Bao, Y.; Xue, Y.; et al. Population Genetics of SARS-CoV-2: Disentangling Effects of Sampling Bias and Infection Clusters. *Genomics, Proteomics Bioinforma.* **2020**. 10.1016/j.gpb.2020.06.001.

[89] Jayamohan, H.; Lambert, C.J.; Sant, H.J.; Jafek, A.; Patel, D.; Feng, H.; Beeman, M.; Mahmood, T.; Nze, U.; Gale, B.K. SARS-CoV-2 pandemic: a review of molecular diagnostic tools including sample collection and commercial response with associated advantages and limitations. *Anal. Bioanal. Chem.* 2021, *413*. 10.1007/s00216-020-02958-1.

[90] Gorzynski, J.E.; de Jong, H.N.; Amar, D.; Hughes, C.R.; Ioannidis, A.; Bierman, R.; Liu, D.; Tanigawa, Y.; Kistler, A.L.; Kamm, J.; et al. High-throughput SARS-CoV-2 and host genome sequencing from single nasopharyngeal swabs. *medRxiv* 2020. 10.1101/2020.07.27.20163147.

[91] Nasir, J.A.; Kozak, R.A.; Aftanas, P.; Raphenya, A.R.; Smith, K.M.; Maguire, F.; Maan, H.; Alruwaili, M.; Banerjee, A.; Mbareche, H.; et al. A comparison of whole genome sequencing of sars-cov-2 using amplicon-based sequencing, random hexamers, and bait capture. *Viruses* **2020**, *12*. 10.3390/v12080895.

[92] Charre, C.; Ginevra, C.; Sabatier, M.; Regue, H.; Destras, G.; Brun, S.; Burfin, G.; Scholtes, C.; Morfin, F.; Valette, M.; et al. Evaluation of NGS-based approaches for SARS-CoV- 2 whole genome characterisation. *Virus Evol.* **2020**, *6*. 10.1093/ve/veaa075.

[93] Gohl, D.M.; Garbe, J.; Grady, P.; Daniel, J.; Watson, R.H.B.; Auch, B.; Nelson, A.; Yohe, S.; Beckman, K.B. A rapid, cost-effective tailed amplicon method for sequencing SARS-CoV-2. *BMC Genomics* **2020**, *21*. 10.1186/s12864-020-07283-6.

[94] Alessandrini, F.; Caucci, S.; Onofri, V.; Melchionda, F.; Tagliabracci, A.; Bagnarelli, P.; Sante, L. Di; Turchi, C.; Menzo, S. Evaluation of the ion ampliseq SARS-CoV-2 research panel by massive parallel sequencing. *Genes (Basel).* **2020**, *11*. 10.3390/genes11080929.

[95] Guo, L.; Boocock, J.; Tome, J.M.; Chandrasekaran, S.; Hilt, E.E.; Zhang, Y.; Sathe, L.; Li, X.; Luo, C.; Kosuri, S.; et al. Rapid cost-effective viral genome sequencing by V-seq. *bioRxiv* 2020. 10.1101/2020.08.15.252510.

[96] Resende, P.C.; Motta, F.C.; Roy, S.; Appolinario, L.; Fabri, A.; Xavier, J.; Harris, K.; Matos, A.R.; Caetano, B.; Orgeswalska, M.; et al. SARS-CoV-2 genomes recovered by long amplicon

tiling multiplex approach using nanopore sequencing and applicable to other sequencing platforms. *bioRxiv* 2020. 10.1101/2020.04.30.069039.

[97] Itokawa, K.; Sekizuka, T.; Hashino, M.; Tanaka, R.; Kuroda, M. A proposal of alternative primers for the ARTIC Network's multiplex PCR to improve coverage of SARS-CoV-2 genome sequencing. *bioRxiv* **2020**.

[98] Li, J.; Wang, H.; Mao, L.; Yu, H.; Yu, X.; Sun, Z.; Qian, X.; Cheng, S.; Chen, S.; Chen, J.; et al. Rapid genomic characterization of SARS-CoV-2 viruses from clinical specimens using nanopore sequencing. *Sci. Rep.* **2020**, *10*. 10.1038/s41598-020-74656-y.

[99] Hourdel, V.; Kwasiborski, A.; Balière, C.; Matheus, S.; Batéjat, C.F.; Manuguerra, J.C.; Vanhomwegen, J.; Caro, V. Rapid Genomic Characterization of SARS-CoV-2 by Direct Amplicon-Based Sequencing Through Comparison of MinION and Illumina iSeq100TM System. *Front. Microbiol.* **2020**, *11*. 10.3389/fmicb.2020.571328.

[100] Fontenele, R.S.; Kraberger, S.; Hadfield, J.; Driver, E.M.; Bowes, D.; Holland, L.A.; Faleye, T.O.C.; Adhikari, S.; Kumar, R.; Inchausti, R.; et al. High-throughput sequencing of SARS-CoV-2 in wastewater provides insights into circulating variants. *medRxiv Prepr. Serv. Heal. Sci.* **2021**. 10.1101/2021.01.22.21250320.

[101] Itokawa, K.; Sekizuka, T.; Hashino, M.; Tanaka, R.; Kuroda, M. Disentangling primer interactions improves SARS-CoV-2 genome sequencing by multiplex tiling PCR. *PLoS One* **2020**, *15*. 10.1371/journal.pone.0239403.

[102] Baker, D.J.; Aydin, A.; Le-Viet, T.; Kay, G.L.; Rudder, S.; de Oliveira Martins, L.; Tedim, A.P.; Kolyva, A.; Diaz, M.; Alikhan, N.F.; et al. CoronaHiT: high-throughput sequencing of SARS-CoV-2 genomes. *Genome Med.* **2021**, *13*. 10.1186/s13073-021-00839-5.

[103] Tyson, J.R.; James, P.; Stoddart, D.; Sparks, N.; Wickenhagen, A.; Hall, G.; Choi, J.H.; Lapointe, H.; Kamelian, K.; Smith, A.D.; et al. Improvements to the ARTIC multiplex PCR method for SARS-CoV-2 genome sequencing using nanopore. *bioRxiv* 2020. 10.1101/2020.09.04.283077.

[104] Wen, S.; Sun, C.; Zheng, H.; Wang, L.; Zhang, H.; Zou, L.; Liu, Z.; Du, P.; Xu, X.; Liang, L.; et al. High-coverage SARS-CoV-2 genome sequences acquired by target capture sequencing. *J. Med. Virol.* **2020**, *92*, 2221-2226. 10.1002/jmv.26116.

[105] Boulware, D.R.; Meya, D.B.; Bergemann, T.L.; Williams, D.; Irina, I.A.; Rhein, J.; Staddon, J.; Kambugu, A.; Janoff, E.N.; Bohjanen, P.R. Antiretroviral therapy down-regulates innate antiviral response genes in patients with AIDS in sub-Saharan Africa. *J. Acquir. Immune Defic. Syndr.* **2010**. 10.1097/QAI.0b013e3181ef4963.

[106] Brienze, V.M.S.; André, J.C.; Liso, E.; Vlasova-St. Louis, I. Cryptococcal immune reconstitution inflammatory syndrome: From blood and cerebrospinal fluid biomarkers to treatment approaches. *Life* 2021, *11*. 10.3390/life11020095.

[107] Vlasova-St. Louis, I.; Chang, C.C.; Shahid, S.; French, M.A.; Bohjanen, P.R. Transcriptomic predictors of paradoxical cryptococcosis-associated immune reconstitution inflammatory syndrome. *Open Forum Infect. Dis.* **2018**, *5*, 1-10. 10.1093/ofid/ofy157.

[108] Mohei, Hesham; Kellampalli, Usha; Vlasova-St. Louis, I.; Vlasova-St. Louis, I. Immune Reconstitution Disorders: Spotlight on Interferons. *Int. J. Biomed. Investig.* **2019**, *2*, 1-21. 10.31531/2581-4745.1000119.

[109] Vlasova-St Louis, I.; Musubire, A.K.; Meya, D.B.; Nabeta, H.W.; Mohei, H.; Boulware, D.R.; Bohjanen, P.R. Transcriptomic biomarker pathways associated with death in HIV-infected patients with cryptococcal meningitis. *BMC Med. Genomics* **2021**, *14*. 10.1186/s12920-021-00914-1.

[110] Zhou, Y.; Fu, B.; Zheng, X.; Wang, D.; Zhao, C.; Qi, Y.; Sun, R.; Tian, Z.; Xu, X.; Wei, H. Pathogenic T-cells and inflammatory monocytes incite inflammatory storms in severe COVID-19 patients. *Natl. Sci. Rev.* 2020, *7*. 10.1093/nsr/nwaa041.

[111] Sawalha, A.H.; Zhao, M.; Coit, P.; Lu, Q. Epigenetic dysregulation of ACE2 and interferon-regulated genes might suggest increased COVID-19 susceptibility and severity in lupus patients. *Clin. Immunol.* **2020**, *215*. 10.1016/j.clim.2020.108410.

[112] Chen, G.; Wu, D.; Guo, W.; Cao, Y.; Huang, D.; Wang, H.; Wang, T.; Zhang, X.; Chen, H.; Yu, H.; et al. Clinical and immunological features of severe and moderate coronavirus disease 2019. *J. Clin. Invest.* **2020**, *130*. 10.1172/JCI137244.

[113] Guo, C.; Li, B.; Ma, H.; Wang, X.; Cai, P.; Yu, Q.; Zhu, L.; Jin, L.; Jiang, C.; Fang, J.; et al. Single-cell analysis of two severe COVID-19 patients reveals a monocyte-associated and tocilizumab-responding cytokine storm. *Nat. Commun.* **2020**, *11*. 10.1038/s41467-020-17834-w.

[114] Leisman, D.E.; Ronner, L.; Pinotti, R.; Taylor, M.D.; Sinha, P.; Calfee, C.S.; Hirayama, A. V.; Mastroiani, F.; Turtle, C.J.; Harhay, M.O.; et al. Cytokine elevation in severe and critical COVID-19: a rapid systematic review, meta-analysis, and comparison with other inflammatory syndromes. *Lancet Respir. Med.* 2020, *8*. 10.1016/S2213-2600(20)30404-5.

[115] Crimi, E.; Benincasa, G.; Figueroa-Marrero, N.; Galdiero, M.; Napoli, C. Epigenetic susceptibility to severe respiratory viral infections and its therapeutic implications: a narrative review. *Br. J. Anaesth.* 2020, *125*. 10.1016/j.bja.2020.06.060.

[116] Saksena, N.; Bonam, S.R.; Miranda-Saksena, M. Epigenetic Lens to Visualize the Severe Acute Respiratory Syndrome Coronavirus-2 (SARS-CoV-2) Infection in COVID-19 Pandemic. *Front. Genet.* **2021**, *12*. 10.3389/fgene.2021.581726.

[117] Herrmann, J.; Adam, E.H.; Notz, Q.; Helmer, P.; Sonntagbauer, M.; Ungemach-Papenberg, P.; Sanns, A.; Zausig, Y.; Steinfeldt, T.; Torje, I.; et al. COVID-19 Induced Acute Respiratory Distress Syndrome—A Multicenter Observational Study. *Front. Med.* **2020**, *7*. 10.3389/fmed.2020.599533.

[118] Aschenbrenner, A.C.; Mouktaroudi, M.; Krämer, B.; Antonakos, N.; Oestreich, M.; Gkizeli, K.; Nuesch-Germano, M.; Saridaki, M.; Bonaguro, L.; Reusch, N.; et al. Disease severity-specific neutrophil signatures in blood transcriptomes stratify COVID-19 patients. *medRxiv* 2020. 10.1101/2020.07.07.20148395.

[119] Cheng, L.C.; Kao, T.J.; Phan, N.N.; Chiao, C.C.; Yen, M.C.; Chen, C.F.; Hung, J.H.; Jiang, J.Z.; Sun, Z.; Wang, C.Y.; et al. Novel signaling pathways regulate SARS-CoV and SARS-CoV-2 infectious disease. *Medicine (Baltimore).* **2021**, *100*. 10.1097/MD.0000000000024321.

[120] Schultze, J.L.; Aschenbrenner, A.C. COVID-19 and the human innate immune system. *Cell* **2021**. 10.1016/j.cell.2021.02.029.

[121] Shah, A. Novel Coronavirus-Induced NLRP3 Inflammasome Activation: A Potential Drug Target in the Treatment of COVID-19. *Front. Immunol.* **2020**, *11*. 10.3389/fimmu.2020.01021.

[122] van den Berg, D.F.; te Velde, A.A. Severe COVID-19: NLRP3 Inflammasome

Dysregulated. *Front. Immunol.* 2020, *11*. 10.3389/fimmu.2020.01580.

[123] Haghjooy Javanmard, S.; Vaseghi, G.; Manteghinejad, A.; Nasirian, M. Neutrophil-to-Lymphocyte ratio as a potential biomarker for disease severity in COVID-19 patients. *J. Glob. Antimicrob. Resist.* **2020**, *22*, 862-863. 10.1016/j.jgar.2020.07.029.

[124] Gianfrancesco, M.; Hyrich, K.L.; Hyrich, K.L.; Al-Adely, S.; Al-Adely, S.; Carmona, L.; Danila, M.I.; Gossec, L.; Gossec, L.; Izadi, Z.; et al. Characteristics associated with hospitalisation for COVID-19 in people with rheumatic disease: Data from the COVID-19 Global Rheumatology Alliance physician-reported registry. *Ann. Rheum. Dis.* **2020**, *79*. 10.1136/annrheumdis-2020-217871.

[125] Sang, E.R.; Tian, Y.; Miller, L.C.; Sang, Y. Epigenetic evolution of ACE2 and IL-6 genes: Non-canonical interferon-stimulated genes correlate to COVID-19 susceptibility in vertebrates. *Genes (Basel).* **2021**, *12*. 10.3390/genes12020154.

[126] Yaqinuddin, A.; Kashir, J. Novel therapeutic targets for SARS-CoV-2-induced acute lung injury: Targeting a potential IL-1β/neutrophil extracellular traps feedback loop. *Med. Hypotheses* **2020**, *143*. 10.1016/j.mehy.2020.109906.

[127] NCT04482699 RAPA-501-Allo Off-the-Shelf Therapy of COVID-19. *https://clinicaltrials.gov/show/NCT04482699* **2020**.

[128] Blanco-Melo, D.; Nilsson-Payant, B.E.; Liu, W.C.; Uhl, S.; Hoagland, D.; Møller, R.; Jordan, T.X.; Oishi, K.; Panis, M.; Sachs, D.; et al. Imbalanced Host Response to SARS-CoV-2 Drives Development of COVID-19. *Cell* **2020**, *181*. 10.1016/j.cell.2020.04.026.

[129] Angioni, R.; Sánchez-Rodríguez, R.; Munari, F.; Bertoldi, N.; Arcidiacono, D.; Cavinato, S.; Marturano, D.; Zaramella, A.; Realdon, S.; Cattelan, A.; et al. Age-severity matched cytokine profiling reveals specific signatures in Covid-19 patients. *Cell Death Dis.* **2020**, *11*. 10.1038/s41419-020-03151-z.

[130] Van Elslande, J.; Vermeersch, P.; Vandervoort, K.; Wawina-Bokalanga, T.; Vanmechelen, B.; Wollants, E.; Laenen, L.; André, E.; Van Ranst, M.; Lagrou, K.; et al. Symptomatic Severe Acute Respiratory Syndrome Coronavirus 2 (SARS-CoV-2) Reinfection by a Phylogenetically Distinct Strain. *Clin. Infect. Dis.* **2020**. 10.1093/cid/ciaa1330.

[131] Tillett, R.L.; Sevinsky, J.R.; Hartley, P.D.; Kerwin, H.; Crawford, N.; Gorzalski, A.; Laverdure, C.; Verma, S.C.; Rossetto, C.C.; Jackson, D.; et al. Genomic evidence for reinfection with SARS-CoV-2: a case study. *Lancet Infect. Dis.* **2021**, *21*, 52-58. 10.1016/S1473-3099(20)30764-7.

[132] K.K.-W., T.; I.F.-N., H.; J.D., I.; A.W.-H., C.; W.-M., C.; A.R., T.; C.H.-Y., F.; S., Y.; H.-W., T.; A.C.-K., N.; et al. COVID-19 re-infection by a phylogenetically distinct SARS-coronavirus-2 strain confirmed by whole genome sequencing. *Clin. Infect. Dis.* **2020**.

[133] Ak, R.; Yilmaz, E.; Seyhan, A.U.; Doganay, F. Recurrence of COVID-19 documented with RT-PCR. *J. Coll. Physicians Surg. Pakistan* **2021**, *31*. 10.29271/jcpsp.2021.Suppl.S26.

[134] Kared, H.; Redd, A.D.; Bloch, E.M.; Bonny, T.S.; Sumatoh, H.R.; Kairi, F.; Carbajo, D.; Abel, B.; Newell, E.W.; Bettinotti, M.; et al. SARS-CoV-2-specific CD8+ T cell responses in convalescent COVID-19 individuals. *J. Clin. Invest.* **2021**. 10.1172/jci145476.

[135] Dearlove, B.; Lewitus, E.; Bai, H.; Li, Y.; Reeves, D.B.; Joyce, M.G.; Scott, P.T.; Amare, M.F.; Vasan, S.; Michael, N.L.; et al. A SARS-CoV-2 vaccine candidate would likely match all

currently circulating variants. *Proc. Natl. Acad. Sci. U. S. A.* **2020**, *117*. 10.1073/pnas.2008281117.

[136] Lipsitch, M.; Grad, Y.H.; Sette, A.; Crotty, S. Cross-reactive memory T cells and herd immunity to SARS-CoV-2. *Nat. Rev. Immunol.* 2020, *20*. 10.1038/s41577-020-00460-4.

[137] Williams, T.C.; Burgers, W.A. SARS-CoV-2 evolution and vaccines: cause for concern? *Lancet Respir. Med.* 2021. 10.1016/S2213-2600(21)00075-8.

[138] The Lancet Respiratory Medicine Realising the potential of SARS-CoV-2 vaccines—a long shot? *Lancet Respir. Med.* 2021, *9*. 10.1016/S2213-2600(21)00045-X.

[139] Hagen Ashley; M.S SARS-CoV-2 Variants vs. Vaccines.

[140] Keehner, J.; Horton, L.E.; Pfeffer, M.A.; Longhurst, C.A.; Schooley, R.T.; Currier, J.S.; Abeles, S.R.; Torriani, F.J. SARS-CoV-2 Infection after Vaccination in Health Care Workers in California. *N. Engl. J. Med.* **2021**. 10.1056/NEJMc2101927.

[141] Kim, J.H.; Marks, F.; Clemens, J.D. Looking beyond COVID-19 vaccine phase 3 trials. *Nat. Med.* 2021, *27*. 10.1038/s41591-021-01230-y.

[142] Fergie, J.; Srivastava, A. Immunity to SARS-CoV-2: Lessons Learned. *Front. Immunol.* **2021**, *12*. 10.3389/fimmu.2021.654165.

[143] Priyanka; Choudhary, O.P.; Singh, I. Evolution of SARS-CoV-2: A prediction on the lineages and vaccine effectiveness. *Travel Med. Infect. Dis.* 2021, *40*. 10.1016/j.tmaid.2021.101983.

[144] Leguia, M.; Vila-Sanjurjo, A.; Chain, P.S.G.; Berry, I.M.; Jarman, R.G.; Pollett, S. Precision Medicine and Precision Public Health in the Era of Pathogen Next-Generation Sequencing. *J. Infect. Dis.* **2020**, *221*. 10.1093/infdis/jiz424.

Chapter 6

Recent Applications of RNA Sequencing in Food and Agriculture

Venkateswara R. Sripathi, Varsha C. Anche, Zachary B. Gossett and Lloyd T. Walker

Abstract

RNA sequencing (RNA-Seq) is the leading, routine, high-throughput, and cost-effective next-generation sequencing (NGS) approach for mapping and quantifying transcriptomes, and determining the transcriptional structure. The transcriptome is a complete collection of transcripts found in a cell or tissue or organism at a given time point or specific developmental or environmental or physiological condition. The emergence and evolution of RNA-Seq chemistries have changed the landscape and the pace of transcriptome research in life sciences over a decade. This chapter introduces RNA-Seq and surveys its recent food and agriculture applications, ranging from differential gene expression, variants calling and detection, allele-specific expression, alternative splicing, alternative polyadenylation site usage, microRNA profiling, circular RNAs, single-cell RNA-Seq, metatranscriptomics, and systems biology. A few popular RNA-Seq databases and analysis tools are also presented for each application. We began to witness the broader impacts of RNA-Seq in addressing complex biological questions in food and agriculture.

Keywords: RNA-Seq, transcriptome, transcripts, genes, variants, gene expression, analysis, applications, databases, and tools

1. Introduction

Transcriptome broadly refers to a collection of RNA transcripts within a particular context that includes combinations of spatial and temporal factors: biological level of organization, from organelle to organism; and phase of growth, differentiation, or development, from zygote through adult. Additionally, one can investigate transcriptomes under more experimental contexts by controlling or varying the factors mentioned above, along with combinations of environmental, genetic, and physiological conditions. All of these factors influence the constituents of a transcriptome, an array of RNA types that traditionally fall into two categories: coding, the messenger RNAs (mRNAs); and non-coding (ncRNAs), such as ribosomal (rRNAs), transfer (tRNA), small interfering (siRNAs), micro (miRNAs), tRNA-derived small (tsRNA), Piwi-interacting (piRNAs), short hairpin (shRNAs), small nuclear (snRNAs), small nucleolar (snoRNAs), long non-coding (lncRNAs), and circular RNAs (circRNAs) [1, 2]. Interestingly, studies have questioned this sharp distinction between coding and non-coding RNAs, paving the way for more research into multifunctional RNA types that transcend this traditional dichotomy [3, 4]. Given the complex definitions of transcriptome and its constituent RNAs, keen attention is required in understanding and managing the context within which a transcriptome is generated and analyzed throughout the experimental procedure and downstream analysis.

Thus far, RNA research efforts have concentrated on a few major types of RNAs: mRNAs, rRNAs, tRNAs, and miRNAs. Accounting for 3-4% of the total RNA in a cell [5], mRNAs are products of transcription and, in eukaryotes, multiple processing steps that usually involve the addition of adenosine monophosphates to form a poly(A) tail via polyadenylation [6]. This coding mRNA is then translated into an amino acid (AA) chain by the ribosome, in a process incorporating ribosomal proteins, AAs, and non-coding RNAs, such as rRNAs and tRNAs. About 60% of the ribosome's mass [7] and up to 95% of the total RNA in a cell [8] can consist of rRNAs, which facilitate mRNA and tRNA binding while catalyzing the transfer of an AA from the tRNA to the growing AA chain. Many processes that comprise gene expression, including the steps mentioned above, can be regulated by miRNAs [9]. These short (17-22 bp), single-stranded, non-coding RNAs are exclusive to eukaryotes and typically bind to complementary sequences on mRNA molecules, thereby inducing degradation or inefficient translation of the target transcript [10].

These four major types of RNA and the multitude of minor types can be selectively isolated and analyzed using various wet lab and dry lab techniques, depending on the specific applications and biological questions under investigation. In the case of transcriptome profiling for coding RNAs in a eukaryotic organism, the ratio of mRNA to rRNAs can be increased: first during library preparation through poly(A) selection, ribosomal depletion, and size selection strategies; and again during the bioinformatic analysis by rRNA filtering during the initial quality control (QC) step in the pipeline. Especially for capturing miRNAs, in addition to rRNA decontamination steps, size selection strategies are used for selective isolation of small RNA [11]. Many bioinformatics tools are available customized for short sequence alignments [12], and a few can evaluate the thermodynamics of miRNA secondary structures [13]. The molecular biology of RNA transcription, processing, transportation, and translation can be drastically different between phylogenetically distant organisms, and hence the taxonomy of the species being studied is often considered. A variety of wet lab and dry lab techniques have been developed to account for the biological differences in mRNA structure and processing throughout the phylogenetic tree of life.

Transcriptome analysis evolved steadily from nucleic acid detection methods (e.g., northern blots), to hybridization-based methods (e.g., microarrays), through a multitude of sequencing-based methods (e.g., RNA-Seq). RNA-Seq has been the most widely used approach for analyzing transcriptomes obtained from phylogenetically diverse organisms [14]. The swift advancements in RNA-Seq research are being driven by the continual improvements in sequencing technologies (first, second, and third generation), which have steadily provided higher throughput, lower cost, and more accurate sequencing for transcriptome analyses. Despite the availability of many sequencing technologies, the Illumina short-read method remains the most widely used platform for transcriptome sequencing, and many consider it as the gold-standard sequencing for single-nucleotide resolution transcriptome analysis with an accuracy of 99.99% and minimal biases [15]. This method has evolved from 35 bp to 350 bp fragment sequencing in the past decade, and it offers multiple library preparation options, including single-end, mate-pair, and paired-end. Library preparation can yield either stranded sequences, where the sense and/or antisense orientation of the output reads is known, or unstranded sequences, where the read orientation is unknown. Stranded RNA-Seq enables the resolution of both sense and antisense transcription for genes overlapping on opposite strands [16], and it remains the standard for most RNA-Seq applications.

A thorough conceptual understanding of the prospective RNA-Seq experiment is required to overcome the plethora of potential biases, errors, misinterpretations, and other various challenges common in RNA-Seq experiments [17, 18]; researchers ought to precisely monitor and engineer each phase of the entire process, wet lab through the dry lab, from beginning to end and in all steps between: experimental design, sample collection, RNA isolation, RNA-QC, adapter ligation, multiplexing, library preparation, library-QC, sequencing, data collection, demultiplexing, pre-processing, data-QC, analyses, and interpretation. The experimental design is the first fundamental process in RNA-Seq analysis. When the goal is to detect statistically significant, differentially expressed genes (DEGs), increasing the number of replicates usually has a more positive effect than increasing the sequencing depth, especially when sequencing over 2 million reads per sample [19, 20]. For most RNA-Seq experiments, six or more biological replicates are recommended, and at least three biological replicates are necessary. If one aims to identify DEGs, then pooling biological replicates before multiplexing is discouraged, but such pooling might be pragmatic when one only attempts to assemble a comprehensive transcriptome. Contrary to biological replicates, technical replicates are unnecessary for RNA-Seq on modern sequencing platforms [19], and resources can be better utilized by increasing the number of biological replicates and minimizing batch effects from unintended influences, such as variance in personnel, in the laboratory environment, and in the selection and usage of materials and methods. A thorough review of the expansive RNA-Seq landscape is available, and to confine our discussion to the scope of this chapter, we will be highlighting the most popular and current RNA-Seq applications in food and agriculture.

2. RNA sequencing (RNA-Seq) applications

2.1 Differential gene expression (DGE)

As previously mentioned, transcriptomes are spatially and temporally dynamic, and they evolve in response to changing environmental, genetic, and physiological conditions. For instance, the transcriptome of one cell type can be significantly different from another cell type, even within the same tissue, and similarly, the transcriptome of a particular cell can vary drastically, as it transitions through the cell cycle, differentiates, acclimates to environmental factors, adapts to the introduction of particular treatments, or changes during disease progression. RNA-Seq can detect such changes in gene expression levels between samples and, in DGE studies, between two or more experimental groups [21, 22]. DGE analysis seeks to identify statistically significant genes that are expressed differently between groups, which are generated through careful attention to experimental design [23]. DGE studies can elucidate functional elements of the genome by identifying gene-level relationships between transcript abundance and experimental conditions, thereby illuminating the mechanisms of associated physiological processes and expanding our understanding of the links between genotype and phenotype [24].

While DGE analysis focuses on quantifying and comparing the complete collection of all transcript isoforms for a gene to identify differentially expressed genes (DEGs), differential isoform expression (DIE) analysis focuses on quantifying and comparing each individual isoform in a collection of transcripts associated with a particular gene, to identify differentially expressed isoforms (DEIs)

Databases			Tools		
S.No	**Database**	**Citation**	**S.No**	**Tool**	**Citation**
1	SPEED2	[26]	1	Ballgown	[27]
2	ImaGEO	[28]	2	limma	[29]
3	GXD	[30]	3	NOISeq	[31]
4	RED	[32]	4	DESeq2	[33]
5	Omnibus database	[34]	5	edgeR	[35]

Table 1.
Some popular databases and tools in finding DEGs in RNA-Seq data.

between experimental groups [25]. The materials and methods for analyzing DEGs differ from those used for DEIs. The decision to find differential genes or isoforms is crucial and determines the downstream analysis, and it is ideally taken at the beginning of the experiment. Given these differences, we discussed the methods most relevant to DGE analysis, since it has been more deeply studied and widely applied. Some methods being applied to investigate DEGs include northern blot, western blot, quantitative real-time PCR (qPCR), expressed sequence tags (ESTs), microarrays, and RNA-Seq. Most bioinformatics pipelines for DGE analysis of RNA-Seq data include five main stages: QC, alignment, quantification, normalization, and DGE calculation, which usually assumes either a negative binomial, log-normal, or nonparametric statistical distribution. Many databases and bioinformatics tools are available for all these stages and downstream analyses, and a few popular, reliable databases and DGE calculation tools are presented below (**Table 1**). Often each program will output slightly different collections of statistically significant DEGs [21], so many investigators use multiple tools, assign higher confidence to intersectional DEGs, and then continue by piping these results through various downstream functional analyses, which will be discussed later in this chapter.

RNA-Seq followed by DGE analysis has been extensively used in the agriculture and food industry. Poultry scientists have applied RNA-Seq analysis to identify DEGs associated with the eggshell formation in the shell gland at different time-points in laying hens [36]. A dairy research group identified significant enrichment of DEGs associated with mammary gland development, milk protein formation, lipid metabolism, and other biological processes linked with milk production traits in lactating cows [37]. Interestingly, the possible roles of DEGs involved in pathogenesis-related pathways in response to peanut allergy have been examined by comparing the transcriptome profiles of high-risk and risk-free infants, facilitating early detection of food allergies in infants [38]. The symbiotic association between rhizobium bacteria and root nodules in leguminous plants is important in agriculture and soil metagenomics, as this interaction improves soil fertility by nitrogen fixation and increases crop production. Differences in nodulation phenotypes have been observed by comparing two diverse symbiotic systems at different time-points using RNA-Seq [39]. Furthermore, these researchers identified DEGs in response to specific strains of rhizobia in soybean roots, and the majority of these DEGs were involved in plant-pathogen interactions and flavonoids biosynthesis [39]. By studying global transcriptome profiles in strawberry fruits, plant scientists have elucidated the influence of red and blue light on the differential expression of genes associated with anthocyanin biosynthesis and accumulation [40].

2.2 Variants calling and detection

The genetic variations in the coding region may or may not alter the amino acid sequence, resulting in asynonymous or synonymous variants, respectively; characterizing such variants is important for associating the genomic locations with a trait or phenotype [41]. RNA-Seq can be used to identify variations in the coding sequences, including single-nucleotide variants (SNVs), short insertions/deletions (indels <50 bp), and structural variants (SVs). SNVs result from a single nucleotide substitution at a particular coordinate and single-nucleotide polymorphism (SNP) refers to a frequent SNV, generally present in at least 1% of the subject population [42]. SNPs are ubiquitous throughout the coding, non-coding, and regulatory regions of the genome. In comparison, a haplotype is a set of genes, alleles, or SNPs, which are inherited together. Copy number variations (CNVs) are a type of SV where regions in the genome are repeated, and the number of these repeats varies among individuals due to duplication or deletion events. The percentage of CNVs detected in diverse organisms varied significantly. Over 80% and > 15% of the detected SNPs and CNVs were associated with gene expression in the mammalian system, respectively [43].

Many experimental methods have been developed to detect genetic variants in the genomes of plants and animals, and a few routinely used techniques include rhAmp (RNase H2-dependent amplification assay), Kompetitive Allele-Specific PCR (KASP), TaqMan, Fluidigm, AmpliSeq, Fluorescence In Situ hybridization (FISH), qRT-PCR, microarray, and RNA-Seq. When generating RNA-Seq data for the downstream bioinformatics analysis, sequencing depth is a major consideration, given its influence on not only the overall results but also the cost of experimentation; and after analyzing variants for mutated myeloid genes, researchers suggested 30-40 million paired-end reads per sample was sufficient [44]. Additionally, highly variable coverage between different genes can hinder variant calling and annotation of RNA-Seq data. To identify variants (SNPs and short indels) in RNA-Seq reads, a typical bioinformatics pipeline involves three phases: data clean-up, variant discovery and filtering, and evaluation. A selection of databases and programs for variant analysis is presented below (**Table 2**).

The application of RNA-Seq in genome-wide screening for genetic variants is imperative to accelerate the usage of genome-based breeding approaches for selecting agriculturally desirable traits in plants [55] and animals [41, 56]. Functional SNPs associated with quality traits (e.g., plant color, flowering, fruit color, size, and ripening) and/or quantitative traits (e.g., grain yield, abiotic, and biotic stress tolerance) may result in phenotypic diversity among individuals. Previous studies have used RNA-Seq analysis to identify SNPs in relatively smaller genomes, such as barley [57], and larger genomes, such as wheat [58]. One of the main

Databases			Tools		
S.No	**Database**	**Citation**	**S.No**	**Tool**	**Citation**
1	AWESOME	[45]	1	AthCNV	[46]
2	KoVariome	[47]	2	GATK workflow	[48]
3	lncRNASNP2	[49]	3	SQUID	[50]
4	SNP2TFBS	[51]	4	DeepVariant	[52]
5	rSNPBase	[53]	5	VarDict	[54]

Table 2.
A few databases and tools in finding structural variants in RNA-Seq.

goals of livestock germplasm improvement is identifying the genetic variation associated with phenotypic traits of economic importance. By screening 15 duck transcriptomes, SNPs in genes related to fat metabolism and digestion were found in genomic regions that have undergone selective pressures [41]. In a similar study, SNPs associated with the fat deposition in sheep have been identified, potentially leading to breeding programs that reduce tail size in fat-tailed phenotypes [59]. While comparing RNA-Seq variant analysis methodologies for investigating beef production in Nellore steers, researchers recently identified SNPs in genes related to feed efficiency, an economically important trait in cattle [60].

2.3 Allele-specific expression (ASE)

RNA-Seq data can be used to investigate allele-specific expressions (ASEs), which denotes a differential expression of two or more alleles in a diploid or a polyploid organism, sometimes may result in multiple traits and phenotypes. Heterozygous SNPs may lead to ASE, and this phenomenon is conserved in most higher organisms, including those in plant and animal kingdoms. Due to the intrinsic potential of heterozygous SNPs, ASE can be a sensitive marker for detecting cis-regulatory variation and reducing background noise in an individual [61]. Heterozygous variants have been identified in coding regions of mRNA, possibly leading to a variant polypeptide or a truncated protein [62]; non-coding regions (splice site, 5'-UTR, or 3'-UTR), possibly influencing mRNA processing and degradation [63]; and non-coding regulatory regions (promoter, enhancer, or silencer), possibly affecting the binding of transcription and epigenetic factors [64]. Genetic and epigenetic factors regulate transcriptional activity and contribute to ASE, and an imbalanced expression via heterozygous SNP loci in a non-haploid genome may lead to a diseased or abnormal condition [65]. Using whole genome sequencing (WGS) alone, variants throughout the entire genome can be identified. However, by combining WGS and RNA-Seq analyses, ASE and allele silencing information can also be obtained.

Of the many bioinformatics tools and databases created to explore ASE, a few are listed here (**Table 3**). However, despite the recent developments in ASE bioinformatics analysis, significant challenges in applying these tools include: 1) required family tree information, i.e., sequencing data from the individual under investigation and their respective parents, which is more laborious and costly; 2) required phased genotype information, i.e., the haplotype of the individual must be known in order to use the source file as input; 3) commonly required genomic and transcriptomic data to obtain ASE, but MBASED (**Table 3**) requires only RNA-Seq data; 4) common usage of short-read data (100-250 bp) due to the low error rate, which is incapable of covering multiple SNVs and subject to read bias at the exon-intron

Databases			Tools		
S.No	Database	Citation	S.No	Tool	Citation
1	dbGaP	[66]	1	EMASE	[67]
2	Genotype-Tissue Expression, GTEx	[68]	2	IDP-ASE	[69]
3	AD ASTRA	[70]	3	QuSAR	[71]
4	dbNSFP	[72]	4	ASEQ	[73]
5	Genevar	[74]	5	MBASED	[75]

Table 3.
Some widely used databases and tools in finding ASE in RNA-Seq data.

junctions; and 5) lack of advanced statistical methods. Long read (1-100 kb) data allows the detection of multiple SNVs, but it is prone to high error rates and low throughput, which is not ideal for downstream ASE quantification. Therefore, researchers can use a hybrid sequencing approach that combines both short and long reads. IDP-ASE (**Table 3**) can utilize such hybrid data to simultaneously phase haplotype and quantify the ASE at both gene and transcript/isoform levels. More sophisticated tools are required to identify ASE associated with multiple phenotypes and complex traits in comprehensive datasets.

Using genome-wide analysis, the underlying genetic and molecular mechanisms associated with ASE in heterosis have been determined in hybrid rice [76]. ASE of *Dof* genes in response to plant hormone signaling and abiotic stresses is likely mediated through cis-regulatory elements that could be useful for sugarcane crop improvement [77]. Genome-wide expression quantitative trait loci (eQTL) and ASE analyses helped identify candidate genes that determine the meat quality traits in pigs [78]. Similarly, ASE is a widespread phenomenon in the bovine genome, and its effects on the meat quality and production traits in Nellore steers have been studied by combining genotyping and RNA-Seq data from skeletal muscle tissue [79]. With RNA-Seq data from three different tissues (liver, fat, and breast muscle) in commercial broiler chickens, researchers examined the biological mechanisms of ASE variants and their associated meat traits in poultry production by using recently developed bioinformatics software, Variant Call Format (VCF) ASE Detection Tool (VADT) [80].

2.4 Alternative splicing (AS)

During the canonical splicing process in eukaryotes, introns are removed as lariats, and the flanking exons are rejoined to form a processed mRNA, with sequences in the RNA determining where splicing occurs. Usually, exons of the same mRNA are spliced, but sometimes exons from different mRNAs can be combined by trans-splicing [81]. The RNA splicing machinery is a complex of proteins called the spliceosome, its major components being small nuclear Ribo-Nuclear Proteins (snRNPs). The three main types of spliceosome complexes are GU–AG spliceosome (major spliceosome), AU–AC spliceosome, and trans-spliceosome [82]. In general, three main classes of RNA splicing are found: pre-mRNA splicing, Group II introns self-splicing, and Group I introns self-splicing. A single gene can produce multiple products by alternative splicing (AS). In addition to normal, canonical splicing, the primary AS events identified in eukaryotes are exon skipping (ES), mutually exclusive exons (EE), alternative 5′ donor sites (A5), alternative 3′ acceptor sites (A3), alternative promoters (AP), intron retention (IR), and alternative polyadenylation (APA) [83]. Of these, the later three events gained attention recently with the advancements in RNA-Seq. AS is often regulated by activator and repressor proteins, and it can lead to premature termination of translation due to the interaction of exon junction complexes (EJC) with release factors, triggering the Nonsense-Mediated mRNA Decay (NMD) pathway [84].

RNA-Seq data can be assembled into full-length isoforms from the raw reads associated with AS of the same gene, and then the corresponding AS events can be identified and characterized. Mate-pair and paired-end sequences have performed better than single-end short-reads for detecting AS patterns [85]. Among the contemporary approaches, long-read sequencing (PacBio/Oxford Nanopore) is an ideal solution for generating full-length transcript sequences and detecting AS events and isoforms [86]. Full-length isoforms can be assembled with or without a reference, and each approach requires specific bioinformatics software. Some of these AS tools and databases are presented here (**Table 4**). Many AS tools can be

Databases			Tools		
S.No	**Database**	**Citation**	**S.No**	**Tool**	**Citation**
1	DIGGER	[87]	1	SplicingFactory	[88]
2	MeDAS	[89]	2	ASpli	[90]
3	ASlive	[91]	3	ASGAL	[92]
4	CuAS	[93]	4	MAJIQ	[94]
5	SpliceDisease	[95]	5	rMATS	[96]

Table 4.
A few popular databases and tools in finding AS events in RNA-Seq data.

used to analyze these AS events genome-wide and/or for a single gene. For example, the ASGAL pipeline (**Table 4**) begins by building a splice graph from a reference genome and an annotation file. Then, the RNA-Seq reads are aligned to the splice graph. Finally, these splice graph alignments are used to detect novel AS events.

Emerging functional roles of AS in generating transcriptomic and proteomic diversity have been evident in diverse biological processes [97]. In the tea leaves of a *Camellia sinensis* cultivar, approximately 64% of genes underwent an AS event, and many of these events were influenced by heat, drought, and their combined stresses [98]. Naturally occurring splice variants in the population have been used in detecting genotype-specific AS events, and in turn, these events have served as biomarkers for genome-wide association studies (GWAS) in rice subjected to salt stress [99]. Comparative transcriptome analyses of fruit, seedling, and flower tissues in tomatoes revealed more AS events in fruits. About 60% of the tomato's multi-exon genes undergo AS events, among which IR is prevalent. Also, the gene expression is preferentially regulated at the isoform level during early fruit development [100].

2.5 Alternative polyadenylation (APA) site usage

During post-transcriptional processing at the 3'UTR region of pre-mRNA, differential usage of polyadenylation sites can lead to a diverse set of transcript isoforms with different 3'UTR lengths and sequences, as part of a ubiquitous regulatory mechanism called Alternative Polyadenylation (APA). Most eukaryotic genes have multiple APA sites (APAs) that are often found in a coding region (CR-APA) or 3'UTR (UTR-APA) [101]. APAs found in internal intronic and exonic regions account for a small proportion of identified APAs, but these predominantly disrupt the coding regions and can result in variable protein isoforms or NMD decay [102]. In contrast, APAs found in the terminal exon and 3'UTR regions account for a significant proportion of identified APAs, and though such APAs usually do not disrupt the coding regions, they may result in transcript isoforms with variable lengths. A poly(A) tail in the 3'UTR region of an mRNA transcript generally provides mRNA stability, localization, and translational efficiency, so these factors are subject to APA-mediated regulation [103]. Since the 3'UTR region can have hotspots for the binding of miRNAs and RNA-binding proteins (RBPs), any modifications in this region may lead to new RNA species interactions or the formation of novel secondary structures, thereby affecting translational efficiency [101, 103]. APAs likely play a role in many processes involved in gene expression, including nuclear export, localization, stability, degradation, repression, translation, and protein diversification [104]. Additionally, APAs associated with differentiation, proliferation, and tissue-specific expression have been reported [105].

Databases			Tools		
S.No	Database	Citation	S.No	Tool	Citation
1	TREND-DB	[106]	1	Deerect-apa	[107]
2	Animal-APAdb	[108]	2	APAlyzer	[109]
3	PlantAPAdb	[110]	3	scDAPA	[111]
4	APAatlas	[112]	4	DeepPASTA	[113]
5	APADB	[114]	5	TAPAS	[115]

Table 5.
Some popular databases and tools in finding APAs in RNA-Seq data.

APAs at the gene-level can be discovered using EST, microarray, RNA-Seq, 3' RNA-Seq, and qRT–PCR methodologies. However, genome-wide screening for APAs can be achieved through NGS based approaches, such as Whole Transcriptome Termini Site sequencing (WTTS-Seq), poly(A) site sequencing (PAS-Seq), direct RNA sequencing (DRS), poly(A) single-molecule sequencing, as well as 3′ region extraction and deep sequencing (3′ READS). Moreover, researchers can engage in cell type-specific APA profiling by preprocessing the samples with specialized wet-lab methods, such as cell sorting, crosslinking immunoprecipitation and green fluorescent protein (GFP)-tagging, and cellular and molecular barcoding. All these methods utilize total RNA or mRNA as their starting material, but they diverge in their usage of polyA enrichment, library preparation, and sequencing strategies. Usually, NGS data analysis for APAs includes preprocessing, size selection, QC, mapping/assembly, normalized expression value assessment for the poly(A) enriched 3'UTRs or transcripts, DGE, functional annotation, motif analysis, and pathway analysis. A few tools that use most of these steps and databases for APA analysis are presented (**Table 5**).

APA processing has been associated with around 70% of human genes, with the longest resulting isoform for each usually observed to be the most abundant [102, 116]. Recent studies have proposed a role for APAs in leaf development and stress response in the two dominant rice (*Oryza sativa* L.) subspecies, indica and japonica, possibly accounting for significant differences in their phylogenetic divergence [117]. They also demonstrated that variations in 3'UTR length from APA resulted in DEGs associated with many important agronomic traits related to rice yield [117]. The possible role of APA in remodeling root-associated transcriptomes has been observed in Sorghum [118], Bamboo [119], and Arabidopsis [120] in response to diverse abiotic stresses. Currently, APA is underexplored and offers many opportunities for significant contributions to the food and agriculture sectors.

2.6 microRNA (miRNA) profiling

RNA-Seq can identify and characterize diverse classes of small (17-200 bp) ncRNAs, including miRNAs, siRNAs, piRNAs, tsRNAs, snoRNAs, and snRNAs. Almost all types of RNAs crosstalk, and especially miRNAs, the abundant class of sRNAs act as mediator molecules in regulating and deregulation of genes via complementary binding to miRNA response elements (MREs) on target transcripts [121]. Moreover, co-localization and co-expression of ncRNA and mRNA and their interactions are well established [122]. MiRNA genes can be found in exonic, intronic, and intergenic regions of the genome, and they are predominantly localized, form clusters, and generally transcribed together as a single transcriptional unit. The various miRNAs can positively and/or negatively regulate gene expression post-transcriptionally or by translational repression [123]. While competing endogenous RNA, ceRNAs (e.g., lncRNAs and circRNAs) contain MREs and can regulate gene expression by acting as “miRNA sponges”, thus reducing the availability of one or more miRNAs for other potential targets [121]. A nascent miRNA transcript undergoes post-transcriptional processing and nuclear export during the canonical regulation, eventually being loaded into the RNA-induced silencing complex (RISC) [124]. After the incorporated miRNA binds to a target mRNA at MREs often located in the 3’-UTR, RISC mediates gene expression by post-transcriptional gene silencing (PTGS) or by mRNA cleavage or mRNA degradation [124]. However, the presence of ceRNAs challenges the canonical miRNA regulation of gene targets, and the mechanisms and functions of miRNA sponges are still unclear [121].

Though several wet lab and computational methods have been evolved in the past two decades for genome-wide screening of miRNAs, in silico approaches, continue to be more widely used due to the ease in exploring the properties of miRNAs. MiRNAs are highly conserved, and the thermodynamics of miRNA secondary structures and target binding have been elucidated; identification of conserved and novel miRNAs and their targets can be performed using readily available bioinformatics tools. A few frequently accessed databases and tools used are listed here (**Table 6**). Most studies have applied homology-based approaches in identifying conserved miRNAs, and miRNA precursors can be identified by conducting secondary structure analysis using RNAfold [140] or mfold [141]. The properties of miRNAs, such as cooperativity and multiplicity, can also predict miRNAs and their targets computationally [123].

Since the first reported miRNAs in *C. elegans*, different miRNAs have been identified in numerous organisms across multiple kingdoms [123]. Several studies have demonstrated their involvement in various biological processes and their potential to alter key agronomic traits [142]. Using RNA-Seq, the functional roles of miRNAs in various stresses (heat, drought, and salinity) have been reported in Arabidopsis [143] and Cotton [144]. Also, many conserved and novel miRNAs and their putative gene targets were identified in Upland cotton and its closest progenitor species using RNA-Seq, and the majority of these targets were transcription factors that were involved in the regulation of fiber growth and development and stress responses [123]. The role of miRNAs in various diseases has been established over two decades, but, recently some naturally occurring food-derived compounds and exogenous diet-derived miRNAs have been implicated in determining the human gut-associated miRNA expression and their profiles, which contributes to human health and well-being of an individual [145].

	Databases		**miRNA gene prediction tools**		**miRNA target prediction tools**	
S.No	**Database**	**Citation**	**Tool**	**Citation**	**Tool**	**Citation**
1	Rfam	[125]	UEA sRNA workbench	[126]	miRWalk	[127]
2	deepBase	[128]	Mirnovo	[129]	mirDIP	[130]
3	miRDB	[131]	miReader	[132]	psRNATarget	[133]
4	miRbase	[134]	miRDeep-star	[135]	TargetScan	[136]
5	Noncode	[137]	miRNAkey	[138]	mirSOM	[139]

Table 6.
A few popular databases and tools for miRNA analysis using RNA-Seq.

2.7 Circular RNAs

Among the many ncRNAs species, circRNAs are characterized by a stable, closed-loop structure formed through back-splicing via an upstream splice acceptor (SA) site, in contrast to the downstream SA sites of standard linear splicing [146]. CircRNAs span exonic, intronic, intergenic regions, UTR (5′ and 3′), and lncRNA loci [147], and they are stable, conserved, non-random, as well as cell-type and tissue-specific [146]. Additionally, circRNAs have been found in all life domains, and, similar to miRNAs, their orthologous expression facilitates discovery, validation, and functional assignments. CircRNAs are transcribed at higher levels than mRNA in specific cells, tissues, or conditions, and they are expressed during chromatin remodeling [146] and in some disease-specific contexts [148]. For example, 14.4% of actively transcribed genes in human fibroblasts produced circRNAs [147], and due to their orthologous, tissue-specific, and spatial expression tendencies, circRNAs may be employed as plausible biomarkers in disease control and treatment [148]. Biological functions for circRNAs continue to be discovered and currently include scaffolding for RNA-binding proteins; formation of regulatory complexes; promotion of translation; regulation of protein function; and target decoys for other regulatory molecules, like miRNAs [149].

Similar to the methods used in experimental validation of linear mRNA, circRNA-forming exons can be determined by RNA-Seq, back-splice junction specific quantitative PCR (qPCR), northern blot, microarrays, RNA fluorescence in situ hybridization (FISH), Chromatin immunoprecipitation (ChIP), RNA immunoprecipitation (RIP), RNA pulldown, mass spectrometry, *in vitro* synthesis, luciferase reporter assays, and denaturing PAGE. RNase-R treated poly(A) mRNA samples and polyadenylated RNA-Seq are ideal for enriching and identifying circRNAs. These circRNAs can also be characterized by utilizing overexpression (cis/trans), knockdown (RNAi machinery), or knockout (CRISPR/Cas9 system) strategies. Based on the presence of a back-splice junction spanning locations in the RNA-Seq reads, researchers can characterize various types of circRNAs in their data [150] with a variety of bioinformatics tools and databases available (**Table 7**).

The biogenesis mechanisms and functional roles of plants are different from animals, but their expression-specific patterns are very similar [161]. Plant circRNAs have been implicated in stress-induced (dehydration, chilling, high-light, etc.) expression patterns [162]. Intricate regulatory roles of circRNAs in ripening through ethylene signaling pathway has been investigated using integrated RNA-Seq and bioinformatics analysis in tomato [163]. The role of circRNAs in the fat deposition by regulating adipogenic differentiation and lipid metabolism has been determined by studying subcutaneous adipose tissues of two pig breeds using

Databases			Tools		
S.No	**Database**	**Citation**	**S.No**	**Tool**	**Citation**
1	Circbank	[151]	1	circRNAprofiler	[152]
2	exoRBase	[153]	2	CircPlant	[154]
3	PlantcircBase	[155]	3	CircCode	[156]
4	circRNADb	[157]	4	Circ RNA wrap	[158]
5	circBase	[159]	5	Circtools	[160]

Table 7.
Some databases and tools in finding circular RNAs from RNA-Seq data.

RNA-Seq and bioinformatics and their potential to serve as early diagnostic markers in treating metabolism-related diseases [164]. CircRNAs found on four casein genes in the bovine mammary gland harbor complementary sites for specific miRNAs, suggesting their regulatory role in milk protein synthesis. These circRNAs can be used to fine-tune the gene expression of casein genes, thus producing high-quality milk protein and enhanced milk in dairy cows [165].

2.8 Single-cell RNA-Seq

Cell-specific transcriptome changes are critical for understanding single cells or groups of cells throughout tissues, organs, and organ systems. Single-cell RNA-Seq (scRNA-Seq) can be used to measure individual gene expression in a single cell and the distribution of expression levels across a cell population. It was first developed to undertake the whole-transcriptome analysis of a single mouse blastomere [166] and gained widespread popularity recently due to sequencing chemistry advancements and the steep decline in sequencing costs since 2014. scRNA-Seq can illuminate the complex interplay between intrinsic cellular processes and extrinsic stimuli in cell fate determination [167], and scRNA-Seq can facilitate novel discovery species or regulatory processes, which may serve as tools in biotechnology and medicine [168]. Many scRNA-Seq protocols have been developed, often differing in their methods used for cell isolation [169], but studies continue to be limited by the difficulties of culturing certain cell types and by issues involving accurate and precise viable cell isolation [170].

Different methodologies are available in generating single-cell RNA-Seq data from a biological sample. However, most of these methodologies utilize these steps: 1) digest the tissue, i.e., single-cell dissociation; 2) isolate single cells by plate-based or droplet-based methods; 3) capture intracellular mRNA and prepare the massively multiplexed library with sample-specific cellular barcodes or unique molecular identifiers (UMI); 4) sequence on an NGS platform to generate raw reads. Several different platforms and frameworks (stand-alone, cloud-based, and interactive web-based) are presently available for conducting the bioinformatics analysis of scRNA-Seq data, and a few examples for each platform are listed in **Table 8**. The majority of scRNA-Seq frameworks partially or fully follow these steps: QC; alignment; mapping QC; cell QC; normalization; batch correction; imputation; cell cycle-assignment; feature selection; dimensionality reduction and visualization; pseudotime; cell type annotation; DGE; unsupervised clustering; and network analysis.

scRNA-Seq has been a valuable tool in determining differential gene expression by using gene cluster analyses among heterogeneous cell types and understanding their complex interactions and cellular responses in woody plants [186]. The use

	Databases		Web-based scRNA-tools		Cloud-based scRNA-tools	
S.No	**Database**	**Citation**	**Tool**	**Citation**	**Tool**	**Citation**
1	SC2disease	[171]	scMappR	[172]	GranatumX	[173]
2	Curated database	[174]	CHARTS	[175]	Cumulus	[176]
3	PanglaoDB	[177]	alona	[178]	SCelVis	[179]
4	scRNA-tools database	[180]	SingleCellNet	[181]	PscB	[182]
5	scRNASeqDB	[183]	Single Cell Explorer	[184]	Falco	[185]

Table 8.
A few popular databases and tools for single-cell RNA-Seq analysis.

of scRNA-Seq and single-cell gene regulatory networks (scGRN) frameworks in studying complex agronomic traits and resistance to various stresses in crops have been proposed [187]. Gene expression profiles among subcellular populations of the skeletal muscle and its development in chicken have been determined using scRNA-Seq, which are important in producing quantity and quality meat in poultry [188]. In sea urchins, using scRNA-Seq, different cell types commonly seen during the embryo development have been identified by the selective inhibition of Delta/Notch and Wnt responsive pathways [189]. Studying the infant and adult cattle mammary glands (MG) with scRNA-Seq, dairy scientists developed a MG-specific single-cell atlas, determined the cell-type heterogeneity, and identified a novel myofibroblast that can differentiate into luminal epithelial cells, and has potential role in lactation and immunity [190].

2.9 Metatranscriptomics

Metatranscriptome refers to the total RNA sequences (protein-coding and non-coding) collected from a location or source or body, which corresponds to the expression profiles of prokaryotic and eukaryotic species found in natural environments such as soil, sea, space, gut, airways, feces, and skin [191]. Metagenomics focuses on the overall genetic composition of the microbial community, while metatranscriptomics provides more profound insights about the genes expressed, their abundance, diversity, differential expression, and aims to address the functional, metabolic, and pathway diversity present in a microbial community [192]. Metatranscriptome is a dynamic entity that can detect gene expression variability with time and environmental changes [193]. Metatranscriptomics is a culture-free profiling method that helps understand the structure (i.e., microbial communities and taxonomic analysis), function (DEGs, enrichment, and annotation), and mechanisms (adaptability, selection, and domestication) of complex microbial communities [194]. It also helps in understanding RNA-mediated regulation and in deriving biological signatures associated with microbial communities.

The experimental methods for analyzing RNA, such as northern blot, qRT-PCR, microarrays, cDNA clone-based Sanger sequencing, and RNA-Seq, are also used for studying and analyzing metatranscriptomes. The main challenges in molecular metatranscriptome methods include low total RNA yield commonly found in environmental samples, high rRNA content in total RNA and its removal, and the fidelity of microbial mRNA isolated. Metatranscriptome analysis using RNA-Seq can distinguish and handle metadata [195], whereas the previous transcript analysis approaches failed to: categorize or catalog metadata, understand community-wide gene expression, and determine functional diversity. Most of the metatranscriptome tools utilize one or more steps from the following: 1) preprocessing (QC, trimming, and filtering), 2) Binning, 3) Mapping or *de novo* assembly, 4) taxonomic units, 5) species profiling, 6) DEGs, 7) annotation and function assignment, and 8) pathway or network analysis [193]. The key challenges in metatranscriptome analysis are: the lack of comprehensive datasets from diverse groups of samples and their associated metadata; the scarcity of metagenomic reference data; the small overlap between metagenome and metatranscriptome datasets; rRNA filtering; and the enrichment of low-abundance mRNAs. Some databases and tools routinely used to access or analyze metatranscriptomes are presented here (**Table 9**).

Though several applications have been documented in the recent past, only selected studies from agriculture and food disciplines are presented here. In agriculture, metatranscriptome analysis can help us find beneficial and harmful rhizosphere-associated microbes specific to plant and soil types. Thus, it allows us to enrich associated rhizosphere microbes that promote crop health and yield.

Databases			Tools		
S.No	Database	Citation	S.No	Tool	Citation
1	SILVA	[196]	1	QIIME 2	[197]
2	Greengenes	[198]	2	SAMSA2	[199]
3	eggNOG database	[200]	3	ASaiM	[201]
4	NCBI RefSeq	[202]	4	MG-RAST	[203]
5	SEED Subsystems	[204]	5	MetaTrans	[205]

Table 9.
Some widely used databases and tools in metatranscriptomics analysis.

Metatranscriptomics has been used in deciphering multifunctional genes and enzymes linked with the degradation of contaminants in the crop rhizosphere [206]. Metatranscriptomic profiling helped to determine the variation in the rumen's microbial composition based on the host feed efficiency in beef cattle [207]. In the food industry, metatranscriptomics can be applied to detect food contamination, toxins, and metabolic activities of food-associated microbes and enhance food safety, quality, and function. Metatranscriptomics has been used in finding insights into the core functional microbiota of soy sauce aroma type liquor production in the fermentation process under varied environmental conditions [208]. Metatranscriptome analysis has been used to study the community dynamics of bacteria in fermented foods [209]. Using metatranscriptome sequencing followed by 16S and 18S rRNA analysis, temperature-induced changes in the structural landscape and functional diversity of the mesophilic and thermophilic food web communities respond to two contrasting temperatures in the rice fields have been observed [210].

2.10 Systems biology/biological network analysis

The ultimate goal of RNA-Seq analysis is to understand the underlying biological processes and mechanisms linked with gene expression and regulation. From molecule to biospheres, biological systems can be represented as networks of pairwise relationships between biological entities throughout various levels of organization. The interactions between biomolecules can be: direct, via physical contact, or indirect, via causal chains or mere correlations. Interactomes that are commonly studied include networks between: DNA–RNA; DNA-Protein; RNA–RNA; RNA-Protein; and Protein–Protein. Theoretically, any network of words can be merged with these interactions, as some elements are shared by both, like common gene, transcript, or protein identifiers. The systems biology approach examines the overall structure and function of a cell or an organism, rather than looking at its components as isolated events [211]. The systems biology approach considers gene expression of an organism or an interaction as a sum of individual genes, sets of genes, and other compounding factors [212]. Gene regulatory networks (GRNs) and co-expression analyses are common elements while studying a biological problem as a system rather than as an individual problem [213].

Given the growing avalanche of RNA-Seq data along with the wealth of network analysis (NA) programs, there are tremendous opportunities to find networks within and between their available datasets, guiding them toward valuable insights, future validation experiments, and a more holistic understanding of their research. NA of RNA-Seq data can illuminate the interrelationships and functional associations [214] between several elements: regulators/co-regulators,

Databases			Tools		
S.No	Database	Citation	S.No	Tool	Citation
1	DualSeqDB	[216]	1	pARACNE	[217]
2	KBase	[218]	2	SCENIC	[219]
3	MODOMICS	[220]	3	SERGIO	[221]
4	EcoCyc	[222]	4	GRNBoost2	[223]
5	doRiNA	[224]	5	dynGENIE3	[225]

Table 10.
A few databases and tools for systems biology analysis using RNA-Seq.

upstream/downstream sequences, and genic features; differentially expressed subnetworks; global connectivity among genes and gene networks. Often combined with the aforementioned biomolecular interactions, a more abstracted view of biological systems can be provided by semantic networks, which involve the relationships between categories of biological meaning, commonly ontological, that have been assigned to the biomolecules. Traditional systems biology relied on mathematical and statistical models. In contrast, modern systems biology depends on computer models that simulate an organism's entire biological systems by considering all components [215]. So, these approaches depend on the constant selection of predictors, building models, and testing. Thus, it allows us to move from descriptive science to data science in providing a holistic answer to the biological question under investigation. Thankfully, the inherent complexity of systems biology is ameliorated by the availability of many open-source tools to reconstruct and visualize networks (a few tools and databases are presented in **Table 10**).

RNA-Seq data from a plant (maize) and a pathogen (*Aspergillus flavus*) interaction has been studied as a system to determine GRNs and co-regulated expression patterns in early processes of infection in imparting resistance to *A. flavus* in maize [226]. Systems biology approach has been utilized in unraveling the complex interactions among transcriptomic, metabolomic, and organoleptic components in tomatoes using MetGenMAP, MapMan, and Cytoscape tools [227]. Also, the role of systems biology in building genome-scale metabolic models (GEMs) for characterizing plant-pathogen (*Phytophthora infestans*) interaction, and disease prevention using cellular localization and network reconstruction tools such as KEGG, LocTree 3, and RAVEN [228]. In the food industry, a systems biology framework, Allergen Peptide Browser that stores and catalogs mass spectrometry data has been used in detecting food allergens such as egg, casein, nuts, gluten, wheat, soy, and fish in food products by employing selected and multiple reaction monitoring approach [229]. Systems biology's role in deciphering underlying common molecular pathways that regulate adipose tissue growth and development in chicken has been determined by examining gene modules, functional enrichment, and network analysis (KEGG, Cytoscape, and WGCNA package) [230].

3. Conclusions

In conclusion, a combination of multi-omic approaches and bioinformatics tools developed to date has unquestionably expanded the scope of RNA-Seq applications and improved our understanding of gene expression data. In addition to the applications discussed in this chapter, fusion gene analysis, RNA editing,

RNA interference, and Epitranscriptomics can also be used to understand novel functions of the gene, complex interactions, and the interplay between coding and non-coding regions during gene regulation. In the near future, we will be able to: sequence transcriptomes from complex environments, study more comprehensive RNA datasets using data science tools, functionally validate predicted genes using gene-editing technologies, which will positively impact the food and agriculture sectors.

Acknowledgements

The authors acknowledge Ms. Shalini P. Etukuri and Dr. Govind Sharma at Alabama A&M University for reviewing this book chapter. Also, authors would like to thank anonymous reviewers and editor for their efforts in improving this book chapter. The authors acknowledge the funding support by the Capacity Building grant #2020-38821-31103 from the USDA National Institute of Food and Agriculture.

Conflict of interest

The authors declare no conflict of interest.

Author details

Venkateswara R. Sripathi*, Varsha C. Anche, Zachary B. Gossett and Lloyd T. Walker
Center for Molecular Biology, Alabama A&M University, Normal, AL, USA

*Address all correspondence to: v.sripathi@aamu.edu

References

[1] Chen H, Shan G. The physiological function of long-noncoding RNAs. Non-coding RNA research. 2020 Sep 17. DOI: 10.1016/j.ncrna.2020.09.003.

[2] Fernandes JC, Acuña SM, Aoki JI, Floeter-Winter LM, Muxel SM. Long non-coding RNAs in the regulation of gene expression: physiology and disease. Non-coding RNA. 2019 Mar;5(1):17. DOI: 10.3390/ncrna5010017

[3] Li J, Liu C. Coding or noncoding, the converging concepts of RNAs. Frontiers in genetics. 2019 May 22;10:496. DOI: 10.3389/fgene.2019.00496

[4] Hubé F, Francastel C. Coding and non-coding RNAs, the frontier has never been so blurred. Frontiers in genetics. 2018 Apr 18;9:140. DOI:10.3389/fgene.2018.00140

[5] Han F, Lillard SJ. In-situ sampling and separation of RNA from individual mammalian cells. Analytical chemistry. 2000 Sep 1;72(17):4073-4079. DOI: 10.1021/ac000428g

[6] Di Giammartino DC, Nishida K, Manley JL. Mechanisms and consequences of alternative polyadenylation. Molecular cell. 2011 Sep 16;43(6):853- DOI: 10.1016/j.molcel.2011.08.017

[7] Gutell RR, Lee JC, Cannone JJ. The accuracy of ribosomal RNA comparative structure models. Current opinion in structural biology. 2002 Jun 1;12(3):301- DOI: 10.1016/S0959-440X(02)00339-1

[8] Peano C, Pietrelli A, Consolandi C, Rossi E, Petiti L, Tagliabue L, De Bellis G, Landini P. An efficient rRNA removal method for RNA sequencing in GC-rich bacteria. Microbial informatics and experimentation. 2013 Dec;3(1):1-1. DOI: 10.1186/2042-5783-3-1

[9] Ha M, Kim VN. Regulation of microRNA biogenesis. Nature reviews Molecular cell biology. 2014 Aug;15(8):509-524. DOI: 10.1038/nrm3838

[10] Huang Y, Shen XJ, Zou Q, Wang SP, Tang SM, Zhang GZ. Biological functions of microRNAs: a review. Journal of physiology and biochemistry. 2011 Mar 1;67(1):129-139. DOI: 10.1007/s13105-010-0050-6

[11] Zhao S, Zhang Y, Gamini R, Zhang B, von Schack D. Evaluation of two main RNA-seq approaches for gene quantification in clinical RNA sequencing: polyA+ selection versus rRNA depletion. Scientific reports. 2018 Mar 19;8(1):1-2. DOI: 10.1038/s41598-018-23226-4

[12] Chen L, Heikkinen L, Wang C, Yang Y, Sun H, Wong G. Trends in the development of miRNA bioinformatics tools. Briefings in bioinformatics. 2019 Sep;20(5):1836-1852. DOI: 10.1093/bib/bby054

[13] Hertel J, Stadler PF. Hairpins in a Haystack: recognizing microRNA precursors in comparative genomics data. Bioinformatics. 2006 Jul 15;22(14):e197-e202. DOI: 10.1093/bioinformatics/btl257

[14] Wang Z, Gerstein M, Snyder M. RNA-Seq: a revolutionary tool for transcriptomics. Nature reviews genetics. 2009 Jan;10(1):57-63. DOI: 10.1038/nrg2484

[15] Tan G, Opitz L, Schlapbach R, Rehrauer H. Long fragments achieve lower base quality in Illumina paired-end sequencing. Scientific reports. 2019 Feb 27;9(1):1-7. DOI: 10.1038/s41598-019-39076-7

[16] Parkhomchuk D, Borodina T, Amstislavskiy V, Banaru M, Hallen L, Krobitsch S, Lehrach H, Soldatov A. Transcriptome analysis by

strand-specific sequencing of complementary DNA. Nucleic acids research. 2009 Oct 1;37(18):e123-. DOI: 10.1093/nar/gkp596

[17] Ozsolak F, Milos PM. RNA sequencing: advances, challenges and opportunities. Nature reviews genetics. 2011 Feb;12(2):87-98. DOI: 10.1038/nrg2934

[18] Lowe R, Shirley N, Bleackley M, Dolan S, Shafee T. Transcriptomics technologies. PLoS computational biology. 2017 May 18;13(5):e1005457. DOI: 10.1371/journal.pcbi.1005457

[19] Liu Y, Zhou J, White KP. RNA-seq differential expression studies: more sequence or more replication?. Bioinformatics. 2014 Feb 1;30(3):301-304. DOI: 10.1093/bioinformatics/btt688

[20] Baccarella A, Williams CR, Parrish JZ, Kim CC. Empirical assessment of the impact of sample number and read depth on RNA-Seq analysis workflow performance. BMC bioinformatics. 2018 Dec;19(1):1-2. DOI: 10.1186/s12859-018-2445-2

[21] Costa-Silva J, Domingues D, Lopes FM. RNA-Seq differential expression analysis: An extended review and a software tool. PloS one. 2017 Dec 21;12(12):e0190152. DOI: 10.1371/journal.pone.0190152

[22] de Jong TV, Moshkin YM, Guryev V. Gene expression variability: the other dimension in transcriptome analysis. Physiological genomics. 2019 May 1;51(5):145-158. DOI: 10.1152/physiolgenomics.00128.2018

[23] Williams AG, Thomas S, Wyman SK, Holloway AK. RNA-seq data: challenges in and recommendations for experimental design and analysis. Current protocols in human genetics. 2014 Oct;83(1):11-DOI: 10.1002/0471142905.hg1113s83

[24] Adriaens ME, Bezzina CR. Genomic approaches for the elucidation of genes and gene networks underlying cardiovascular traits. Biophysical reviews. 2018 Aug;10(4):1053-1060. DOI: 10.1007/s12551-018-0435-2

[25] Merino GA, Conesa A, Fernández EA. A benchmarking of workflows for detecting differential splicing and differential expression at isoform level in human RNA-seq studies. Briefings in bioinformatics. 2019 Mar;20(2):471-481. DOI: 10.1093/bib/bbx122

[26] Rydenfelt M, Klinger B, Klünemann M, Blüthgen N. SPEED2: inferring upstream pathway activity from differential gene expression. Nucleic acids research. 2020 Jul 2;48(W1):W307-W312. DOI: 10.1093/nar/gkaa236

[27] Frazee AC, Pertea G, Jaffe AE, Langmead B, Salzberg SL, Leek JT. Ballgown bridges the gap between transcriptome assembly and expression analysis. Nature biotechnology. 2015 Mar;33(3):243-246. DOI: 10.1038/nbt.3172

[28] Toro-Domínguez D, Martorell-Marugán J, López-Domínguez R, García-Moreno A, González-Rumayor V, Alarcón-Riquelme ME, Carmona-Sáez P. ImaGEO: integrative gene expression meta-analysis from GEO database. Bioinformatics. 2019 Mar 1;35(5):880-882. DOI: 10.1093/bioinformatics/bty721

[29] Ritchie ME, Phipson B, Wu DI, Hu Y, Law CW, Shi W, Smyth GK. limma powers differential expression analyses for RNA-sequencing and microarray studies. Nucleic acids research. 2015 Apr 20;43(7):e47-. DOI: 10.1093/nar/gkv007

[30] Smith CM, Hayamizu TF, Finger JH, Bello SM, McCright IJ, Xu J,

Baldarelli RM, Beal JS, Campbell J, Corbani LE, Frost PJ. The mouse gene expression database (GXD): 2019 update. Nucleic acids research. 2019 Jan 8;47(D1):D774-D779. DOI: 10.1093/nar/gky922

[31] Tarazona S, Furió-Tarí P, Turrà D, Pietro AD, Nueda MJ, Ferrer A, Conesa A. Data quality aware analysis of differential expression in RNA-seq with NOISeq R/Bioc package. Nucleic acids research. 2015 Dec 2;43(21):e140-. DOI: 10.1093/nar/gkv711

[32] Xia L, Zou D, Sang J, Xu X, Yin H, Li M, Wu S, Hu S, Hao L, Zhang Z. Rice Expression Database (RED): An integrated RNA-Seq-derived gene expression database for rice. Journal of Genetics and Genomics. 2017 May 20;44(5):235-241. DOI: 10.1016/j.jgg.2017.05.003

[33] Love MI, Huber W, Anders S. Moderated estimation of fold change and dispersion for RNA-seq data with DESeq2. Genome biology. 2014 Dec;15(12):1-21. DOI: 10.1186/s13059-014-0550-8

[34] Clough E, Barrett T. The gene expression omnibus database. InStatistical genomics 2016 (pp. 93-110). Humana Press, New York, NY. DOI: DOI: 10.1007/978-1-4939-3578-9_5.

[35] Chen Y, Lun AT, Smyth GK. Differential expression analysis of complex RNA-seq experiments using edgeR. Statistical analysis of next generation sequencing data. 2014:51-74. DOI: 10.1007/978-3-319-07212-8_3

[36] Khan S, Wu SB, Roberts J. RNA-sequencing analysis of shell gland shows differences in gene expression profile at two time-points of eggshell formation in laying chickens. BMC genomics. 2019 Dec;20(1):1-20. DOI: 10.1186/s12864-019-5460-4

[37] Yang J, Jiang J, Liu X, Wang H, Guo G, Zhang Q, Jiang L. Differential expression of genes in milk of dairy cattle during lactation. Animal genetics. 2016 Apr;47(2):174-180. DOI: 10.1111/age.12394

[38] Devonshire AL, Gursel DB, Fan H, Erickson KA, Pongracic JA, Singh AM, Kumar R. Differential Gene Expression Among Infants at High-Risk for Peanut Allergy. Journal of Allergy and Clinical Immunology. 2019 Feb 1;143(2):AB82. DOI: 10.1016/j.jaci.2018.12.255

[39] Yuan S, Li R, Chen S, Chen H, Zhang C, Chen L, Hao Q, Shan Z, Yang Z, Qiu D, Zhang X. RNA-Seq analysis of differential gene expression responding to different rhizobium strains in soybean (Glycine max) roots. Frontiers in plant science. 2016 May 30;7:721. DOI: 10.3389/fpls.2016.00721

[40] Zhang Y, Jiang L, Li Y, Chen Q, Ye Y, Zhang Y, Luo Y, Sun B, Wang X, Tang H. Effect of red and blue light on anthocyanin accumulation and differential gene expression in strawberry (Fragaria× ananassa). Molecules. 2018 Apr;23(4):820. DOI: 10.3390/molecules23040820

[41] Lin R, Du X, Peng S, Yang L, Ma Y, Gong Y, Li S. Discovering all transcriptome single-nucleotide polymorphisms and scanning for selection signatures in ducks (*Anas platyrhynchos*). Evolutionary Bioinformatics. 2015 Jan;11:EBO-S21545. DOI: 10.4137/EBO.S21545

[42] Maughan PJ, Yourstone SM, Byers RL, Smith SM, Udall JA. Single-Nucleotide Polymorphism Genotyping in Mapping Populations via Genomic Reduction and Next-Generation Sequencing: Proof of Concept. The Plant Genome. 2010 Nov;3(3). DOI: 10.3835/plantgenome2010.07.0016

[43] Stranger BE, Forrest MS, Dunning M, Ingle CE, Beazley C,

Thorne N, Redon R, Bird CP, De Grassi A, Lee C, Tyler-Smith C. Relative impact of nucleotide and copy number variation on gene expression phenotypes. Science. 2007 Feb 9;315(5813):848-853. DOI: 10.1126/science.1136678

[44] Quaglieri A, Flensburg C, Speed TP, Majewski IJ. Finding a suitable library size to call variants in RNA-seq. BMC bioinformatics. 2020 Dec;21(1):1-9. DOI: 10.1186/s12859-020-03860-4

[45] Yang Y, Peng X, Ying P, Tian J, Li J, Ke J, Zhu Y, Gong Y, Zou D, Yang N, Wang X. AWESOME: a database of SNPs that affect protein post-translational modifications. Nucleic acids research. 2019 Jan 8;47(D1):D874-D880. DOI: 10.1093/nar/gky821

[46] Zmienko A, Marszalek-Zenczak M, Wojciechowski P, Samelak-Czajka A, Luczak M, Kozlowski P, Karlowski WM, Figlerowicz M. AthCNV: A map of DNA copy number variations in the Arabidopsis genome. The Plant Cell. 2020 Jun 1;32(6):1797-1819. DOI: 10.1105/tpc.19.00640

[47] Kim J, Weber JA, Jho S, Jang J, Jun J, Cho YS, Kim HM, Kim H, Kim Y, Chung O, Kim CG. KoVariome: Korean National Standard Reference Variome database of whole genomes with comprehensive SNV, indel, CNV, and SV analyses. Scientific reports. 2018 Apr 4;8(1):1-4. DOI: 10.1038/s41598-018-23837-x

[48] Brouard JS, Schenkel F, Marete A, Bissonnette N. The GATK joint genotyping workflow is appropriate for calling variants in RNA-seq experiments. Journal of animal science and biotechnology. 2019 Dec;10(1):1-6. DOI: 10.1186/s40104019-0359-0

[49] Miao YR, Liu W, Zhang Q, Guo AY. lncRNASNP2: an updated database of functional SNPs and mutations in human and mouse lncRNAs. Nucleic acids research. 2018 Jan 4;46(D1):D276-D280. DOI: 10.1093/nar/gkx1004

[50] Ma C, Shao M, Kingsford C. SQUID: transcriptomic structural variation detection from RNA-seq. Genome biology. 2018 Dec 1;19(1):52. DOI: 10.1186/s13059-018-1421-5

[51] Kumar S, Ambrosini G, Bucher P. SNP2TFBS–a database of regulatory SNPs affecting predicted transcription factor binding site affinity. Nucleic acids research. 2017 Jan 4;45(D1):D139-D144. DOI: 10.1093/nar/gkw1064

[52] Poplin R, Chang PC, Alexander D, Schwartz S, Colthurst T, Ku A, Newburger D, Dijamco J, Nguyen N, Afshar PT, Gross SS. A universal SNP and small-indel variant caller using deep neural networks. Nature biotechnology. 2018 Nov;36(10):983-987. DOI: 10.1038/nbt.4235

[53] Guo L, Du Y, Chang S, Zhang K, Wang J. rSNPBase: a database for curated regulatory SNPs. Nucleic Acids Research. 2014 Jan 1;42(D1):D1033-D1039. DOI: 10.1093/nar/gkt1167

[54] Lai Z, Markovets A, Ahdesmaki M, Chapman B, Hofmann O, McEwen R, Johnson J, Dougherty B, Barrett JC, Dry JR. VarDict: a novel and versatile variant caller for next-generation sequencing in cancer research. Nucleic acids research. 2016 Jun 20;44(11):e108-. DOI: 10.1093/nar/gkw227

[55] Morgil H, Gercek YC, Tulum I. Single nucleotide polymorphisms (SNPs) in plant genetics and breeding. InThe Recent Topics in Genetic Polymorphisms 2020 Mar 28. IntechOpen. DOI: 10.5772/intechopen.91886

[56] Fang L, Sahana G, Su G, Yu Y, Zhang S, Lund MS, Sørensen P.

Integrating sequence-based GWAS and RNA-Seq provides novel insights into the genetic basis of mastitis and milk production in dairy cattle. Scientific reports. 2017 Mar 30;7(1):1-6. DOI: 10.1038/srep45560

[57] Tanaka T, Ishikawa G, Ogiso-Tanaka E, Yanagisawa T, Sato K. Development of genome-wide SNP markers for barley via reference-based RNA-Seq analysis. Frontiers in plant science. 2019 May10;10:577. DOI: 10.3389/fpls.2019.00577

[58] Nishijima R, Yoshida K, Motoi Y, Sato K, Takumi S. Genome-wide identification of novel genetic markers from RNA sequencing assembly of diverse Aegilops tauschii accessions. Molecular Genetics and Genomics. 2016 Aug;291(4):1681-1694. DOI: 10.1007/s00438-016-1211-2

[59] Bakhtiarizadeh MR, Alamouti AA. RNA-Seq based genetic variant discovery provides new insights into controlling fat deposition in the tail of sheep. Scientific Reports. 2020 Aug 11;10(1):1-3. DOI: 10.1038/s41598-020-70527-8

[60] Lam S, Zeidan J, Miglior F, Suárez-Vega A, Gómez-Redondo I, Fonseca PA, Guan LL, Waters S, Cánovas A. Development and comparison of RNA-sequencing pipelines for more accurate SNP identification: practical example of functional SNP detection associated with feed efficiency in Nellore beef cattle. BMC genomics. 2020 Dec;21(1):1-7. DOI: 10.1186/s12864-020-07107-7

[61] Pastinen T. Genome-wide allele-specific analysis: insights into regulatory variation. Nature Reviews Genetics. 2010 Aug;11(8):533-538. DOI: 10.1038/nrg2815

[62] Kukurba KR, Zhang R, Li X, Smith KS, Knowles DA, Tan MH, Piskol R, Lek M, Snyder M, MacArthur DG, Li JB. Allelic expression of deleterious protein-coding variants across human tissues. PLoS Genet. 2014 May 1;10(5):e1004304. DOI: 10.1371/journal.pgen.1004304

[63] Li G, Bahn JH, Lee JH, Peng G, Chen Z, Nelson SF, Xiao X. Identification of allele-specific alternative mRNA processing via transcriptome sequencing. Nucleic acids research. 2012 Jul 1;40(13):e104-. DOI: 10.1093/nar/gks280

[64] Reddy TE, Gertz J, Pauli F, Kucera KS, Varley KE, Newberry KM, Marinov GK, Mortazavi A, Williams BA, Song L, Crawford GE. Effects of sequence variation on differential allelic transcription factor occupancy and gene expression. Genome research. 2012 May 1;22(5):860-869. DOI: 10.1101/gr.131201.111

[65] Berger E, Yorukoglu D, Zhang L, Nyquist SK, Shalek AK, Kellis M, Numanagić I, Berger B. Improved haplotype inference by exploiting long-range linking and allelic imbalance in RNA-seq datasets. Nature communications. 2020 Sep 16;11(1):1-9. DOI: 10.1038/s41467-020-18320-z

[66] Tryka KA, Hao L, Sturcke A, Jin Y, Wang ZY, Ziyabari L, Lee M, Popova N, Sharopova N, Kimura M, Feolo M. NCBI's Database of Genotypes and Phenotypes: dbGaP. Nucleic acids research. 2014 Jan 1;42(D1):D975-D979. DOI: 10.1093/nar/gkt1211

[67] Raghupathy N, Choi K, Vincent MJ, Beane GL, Sheppard KS, Munger SC, Korstanje R, Pardo-Manual de Villena F, Churchill GA. Hierarchical analysis of RNA-seq reads improves the accuracy of allele-specific expression. Bioinformatics. 2018 Jul 1;34(13):2177-2184. DOI: 10.1093/bioinformatics/bty078

[68] Stanfill AG, Cao X. Enhancing Research Through the Use of the Genotype-Tissue Expression (GTEx) Database. Biological Research For Nursing. 2021 Feb 18:1099800421994186. DOI: 10.1177/1099800421994186

[69] Deonovic B, Wang Y, Weirather J, Wang XJ, Au KF. IDP-ASE: haplotyping and quantifying allele-specific expression at the gene and gene isoform level by hybrid sequencing. Nucleic acids research. 2017 Mar 17;45(5):e32-. DOI: 10.1093/nar/gkw1076

[70] Abramov S, Baulin E, Makeev VJ, Boytsov A, Yevshin I, Kulakovskiy IV, Bykova D, Kolpakov F. AD ASTRA: the database of Allelic Dosage-corrected Allele-Specific TRAnscription factor binding suggests causal regulatory sequence variants of pathologies. InBioinformatics of Genome Regulation and Structure/Systems Biology (BGRS/SB-2020) 2020 (pp. 14-14). DOI: 10.18699/BGRS/SB-2020-001

[71] Harvey CT, Moyerbrailean GA, Davis GO, Wen X, Luca F, Pique-Regi R. QuASAR: quantitative allele-specific analysis of reads. Bioinformatics. 2015 Apr 15;31(8):1235-1242. DOI: 10.1093/bioinformatics/btu802

[72] Liu X, Wu C, Li C, Boerwinkle E. dbNSFP v3. 0: A one-stop database of functional predictions and annotations for human nonsynonymous and splice-site SNVs. Human mutation. 2016 Mar;37(3):235-241. DOI: 10.1002/humu.22932

[73] Romanel A, Lago S, Prandi D, Sboner A, Demichelis F. ASEQ: fast allele-specific studies from next-generation sequencing data. BMC medical genomics. 2015 Dec;8(1):1-2. DOI: 10.1186/s12920-015-0084-2

[74] Yang TP, Beazley C, Montgomery SB, Dimas AS, Gutierrez-Arcelus M, Stranger BE, Deloukas P, Dermitzakis ET. Genevar: a database and Java application for the analysis and visualization of SNP-gene associations in eQTL studies. Bioinformatics. 2010 Oct 1;26(19):2474-2476. DOI: 10.1093/bioinformatics/btq452

[75] Mayba O, Gilbert HN, Liu J, Haverty PM, Jhunjhunwala S, Jiang Z, Watanabe C, Zhang Z. MBASED: allele-specific expression detection in cancer tissues and cell lines. Genome biology. 2014 Aug;15(8):1-21. DOI: 10.1186/s13059-014-0405-3

[76] Shao L, Xing F, Xu C, Zhang Q, Che J, Wang X, Song J, Li X, Xiao J, Chen LL, Ouyang Y. Patterns of genome-wide allele-specific expression in hybrid rice and the implications on the genetic basis of heterosis. Proceedings of the National Academy of Sciences. 2019 Mar 19;116(12):5653-5658. DOI: 10.1073/pnas.1820513116

[77] Cai M, Lin J, Li Z, Lin Z, Ma Y, Wang Y, Ming R. Allele specific expression of Dof genes responding to hormones and abiotic stresses in sugarcane. PloS one. 2020 Jan 16;15(1):e0227716. DOI: 10.1371/journal.pone.0227716

[78] Liu Y, Liu X, Zheng Z, Ma T, Liu Y, Long H, Cheng H, Fang M, Gong J, Li X, Zhao S. Genome-wide analysis of expression QTL (eQTL) and allele-specific expression (ASE) in pig muscle identifies candidate genes for meat quality traits. Genetics Selection Evolution. 2020 Dec;52(1):1-1. DOI: 10.1186/s12711-020-00579-x

[79] de Souza MM, Zerlotini A, Rocha MI, Bruscadin JJ, da Silva Diniz WJ, Cardoso TF, Cesar AS, Afonso J, Andrade BG, de Alvarenga Mudadu M, Mokry FB. Allele-specific expression is widespread in Bos indicus muscle and affects meat quality candidate genes. Scientific Reports. 2020 Jun 23;10(1):1-1. DOI: 10.1038/s41598-020-67089-0

[80] Tomlinson MJ, Polson SW, Qiu J, Lake JA, Lee W, Abasht B. Investigation of allele specific expression in various tissues of broiler chickens using the detection tool VADT. Scientific reports. 2021 Feb 17;11(1):1-3. DOI: 10.1038/s41598-021-83459-8

[81] Reynolds DJ, Hertel KJ. Ultra-deep sequencing reveals pre-mRNA splicing as a sequence driven high-fidelity process. PloS one. 2019 Oct 3;14(10):e0223132 DOI: 10.1371/journal.pone.0223132.

[82] Will CL, Lührmann R. Spliceosome structure and function. Cold Spring Harbor perspectives in biology. 2011 Jul 1;3(7):a003707. DOI:10.1101/cshperspect.a003707

[83] Hu H, Yang W, Zheng Z, Niu Z, Yang Y, Wan D, Liu J, Ma T. Analysis of alternative splicing and alternative polyadenylation in *Populus alba var. pyramidalis* by single-molecular long-read sequencing. Frontiers in genetics. 2020 Feb 7;11:48. DOI: 10.3389/fgene.2020.00048

[84] Karousis ED, Nasif S, Mühlemann O. Nonsense-mediated mRNA decay: novel mechanistic insights and biological impact. Wiley Interdisciplinary Reviews: RNA. 2016 Sep;7(5):661-682. DOI: 10.1002/wrna.1357

[85] Rossell D, Attolini CS, Kroiss M, Stöcker A. Quantifying alternative splicing from paired-end RNA-sequencing data. The annals of applied statistics. 2014 Mar;8(1):309. DOI: 10.1214/13-aoas687

[86] Amarasinghe SL, Su S, Dong X, Zappia L, Ritchie ME, Gouil Q. Opportunities and challenges in long-read sequencing data analysis. Genome biology. 2020 Dec;21(1):1-6. DOI: 10.1186/s13059-020-1935-5

[87] Louadi Z, Yuan K, Gress A, Tsoy O, Kalinina OV, Baumbach J, Kacprowski T, List M. DIGGER: exploring the functional role of alternative splicing in protein interactions. Nucleic Acids Research. 2021 Jan 8;49(D1):D309-D318. DOI: 10.1093/nar/gkaa768

[88] Szikora P, Pór T, Sebestyen E. SplicingFactory-Splicing diversity analysis for transcriptome data. bioRxiv. 2021 Jan 1. DOI: 10.1101/2021.02.03.429568

[89] Li Z, Zhang Y, Bush SJ, Tang C, Chen L, Zhang D, Urrutia AO, Lin JW, Chen L. MeDAS: a Metazoan Developmental Alternative Splicing database. Nucleic Acids Research. 2021 Jan 8;49(D1):D144-D150. DOI: 10.1093/nar/gkaa886

[90] Estefania M, Andres R, Javier I, Marcelo Y, Ariel C. ASpli: Integrative analysis of splicing landscapes through RNA-Seq assays. Bioinformatics. 2021 Mar 2. DOI: 10.1093/bioinformatics/btab141

[91] Liu J, Tan S, Huang S, Huang W. ASlive: a database for alternative splicing atlas in livestock animals. BMC genomics. 2020 Dec;21(1):1-7. DOI: 10.1186/s12864-020-6472-9

[92] Denti L, Rizzi R, Beretta S, Della Vedova G, Previtali M, Bonizzoni P. ASGAL: aligning RNA-Seq data to a splicing graph to detect novel alternative splicing events. BMC bioinformatics. 2018 Dec;19(1):1-21. DOI: 10.1186/s12859-018-2436-3

[93] Sun Y, Zhang Q, Liu B, Lin K, Zhang Z, Pang E. CuAS: a database of annotated transcripts generated by alternative splicing in cucumbers. BMC plant biology. 2020 Dec;20(1):1-7. DOI: 10.1186/s12870-020-2312-y

[94] Vaquero-Garcia J, Barrera A, Gazzara MR, Gonzalez-Vallinas J, Lahens NF, Hogenesch JB, Lynch KW, Barash Y. A new view of transcriptome complexity and regulation through the

lens of local splicing variations. elife. 2016 Feb 1;5:e11752. DOI: 10.7554/eLife.11752

[95] Wang J, Zhang J, Li K, Zhao W, Cui Q. SpliceDisease database: linking RNA splicing and disease. Nucleic acids research. 2012 Jan 1;40(D1):D1055-D1059. DOI: 10.1093/nar/gkr1171

[96] Shen S, Park JW, Lu ZX, Lin L, Henry MD, Wu YN, Zhou Q, Xing Y. rMATS: robust and flexible detection of differential alternative splicing from replicate RNA-Seq data. Proceedings of the National Academy of Sciences. 2014 Dec 23;111(51):E5593-E5601. DOI: 10.1073/pnas.1419161111

[97] Wang Y, Liu J, Huang BO, Xu YM, Li J, Huang LF, Lin J, Zhang J, Min QH, Yang WM, Wang XZ. Mechanism of alternative splicing and its regulation. Biomedical reports. 2015 Mar 1;3(2):152-158. DOI: 10.3892/br.2014.407

[98] Ding Y, Wang Y, Qiu C, Qian W, Xie H, Ding Z. Alternative splicing in tea plants was extensively triggered by drought, heat and their combined stresses. PeerJ. 2020 Jan 29;8:e8258. DOI: 10.7717/peerj.8258

[99] Yu H, Du Q, Campbell M, Yu B, Walia H, Zhang C. Genome-wide discovery of natural variation in pre-mRNA splicing and prioritising causal alternative splicing to salt stress response in rice. New Phytologist. 2021 Jan 16. DOI: 10.1111/nph.17189

[100] Sun Y, Xiao H. Identification of alternative splicing events by RNA sequencing in early growth tomato fruits. BMC genomics. 2015 Dec;16(1):1-3. DOI: 10.1186/s12864-015-2128-6

[101] Chen W, Jia Q, Song Y, Fu H, Wei G, Ni T. Alternative polyadenylation: methods, findings, and impacts. Genomics, proteomics & bioinformatics. 2017 Oct 1;15(5):287-300. DOI: 10.1016/j.gpb.2017.06.001

[102] Tian B, Manley JL. Alternative polyadenylation of mRNA precursors. Nature reviews Molecular cell biology. 2017 Jan;18(1):18-30. DOI: 10.1038/nrm.2016.116

[103] Mayr C. Evolution and biological roles of alternative 3′ UTRs. Trends in cell biology. 2016 Mar 1;26(3):227-237. DOI: 10.1016/j.tcb.2015.10.012

[104] Zhang Y, Liu L, Qiu Q, Zhou Q, Ding J, Lu Y, Liu P. Alternative polyadenylation: methods, mechanism, function, and role in cancer. Journal of Experimental & Clinical Cancer Research. 2021 Dec;40(1):1-9. DOI: 10.1186/s13046-021-01852-7

[105] Li Y, Schaefke B, Zou X, Zhang M, Heyd F, Sun W, Zhang B, Li G, Liang W, He Y, Zhou J. Pan-tissue analysis of allelic alternative polyadenylation suggests widespread functional regulation. Molecular systems biology. 2020 Apr;16(4):e9367. DOI: 10.15252/msb.20199367

[106] Marini F, Scherzinger D, Danckwardt S. TREND-DB—a transcriptome-wide atlas of the dynamic landscape of alternative polyadenylation. Nucleic Acids Research. 2021 Jan 8;49(D1):D243-D253. DOI: 10.1093/nar/gkaa722

[107] Li Z, Li Y, Zhang B, Li Y, Long Y, Zhou J, Zou X, Zhang M, Hu Y, Chen W, Gao X. Deerect-apa: Prediction of alternative polyadenylation site usage through deep learning. Genomics, Proteomics & Bioinformatics. 2021 Mar 2. DOI: 10.1016/j.gpb.2020.05.004

[108] Jin W, Zhu Q, Yang Y, Yang W, Wang D, Yang J, Niu X, Yu D, Gong J. Animal-APAdb: a comprehensive animal alternative polyadenylation database. Nucleic Acids Research. 2021 Jan

8;49(D1):D47-D54. DOI: 10.1093/nar/gkaa778

[109] Wang R, Tian B. APAlyzer: a bioinformatics package for analysis of alternative polyadenylation isoforms. Bioinformatics. 2020 Jun 1;36(12):3907-3909. DOI: 10.1093/bioinformatics/btaa266

[110] Zhu S, Ye W, Ye L, Fu H, Ye C, Xiao X, Ji Y, Lin W, Ji G, Wu X. PlantAPAdb: a comprehensive database for alternative polyadenylation sites in plants. Plant physiology. 2020 Jan 1;182(1):228-242. DOI: 10.1104/pp.19.00943

[111] Ye C, Zhou Q, Wu X, Yu C, Ji G, Saban DR, Li QQ. scDAPA: detection and visualization of dynamic alternative polyadenylation from single cell RNA-seq data. Bioinformatics. 2020 Feb 15;36(4):1262-1264. DOI: 10.1093/bioinformatics/btz701

[112] Hong W, Ruan H, Zhang Z, Ye Y, Liu Y, Li S, Jing Y, Zhang H, Diao L, Liang H, Han L. APAatlas: decoding alternative polyadenylation across human tissues. Nucleic acids research. 2020 Jan 8;48(D1):D34-D39. DOI: 10.1093/nar/gkz876

[113] Arefeen A, Xiao X, Jiang T. DeepPASTA: deep neural network based polyadenylation site analysis. Bioinformatics. 2019 Nov 1;35(22):4577-4585. DOI: 10.1093/bioinformatics/btz283

[114] Müller S, Rycak L, Afonso-Grunz F, Winter P, Zawada AM, Damrath E, Scheider J, Schmäh J, Koch I, Kahl G, Rotter B. APADB: a database for alternative polyadenylation and microRNA regulation events. Database. 2014 Jan 1;2014. DOI: 10.1093/database/bau076

[115] Arefeen A, Liu J, Xiao X, Jiang T. TAPAS: tool for alternative polyadenylation site analysis. Bioinformatics. 2018 Aug 1;34(15):2521-2529. DOI: 10.1093/bioinformatics/bty110

[116] Derti A, Garrett-Engele P, MacIsaac KD, Stevens RC, Sriram S, Chen R, Rohl CA, Johnson JM, Babak T. A quantitative atlas of polyadenylation in five mammals. Genome research. 2012 Jun 1;22(6):1173-1183. DOI: 10.1101/gr.132563.111

[117] Zhou Q, Fu H, Yang D, Ye C, Zhu S, Lin J, Ye W, Ji G, Ye X, Wu X, Li QQ. Differential alternative polyadenylation contributes to the developmental divergence between two rice subspecies, japonica and indica. The Plant Journal. 2019 Apr;98(2):260-276. DOI: 10.1111/tpj.14209

[118] Chakrabarti M, de Lorenzo L, Abdel-Ghany SE, Reddy AS, Hunt AG. Wide-ranging transcriptome remodelling mediated by alternative polyadenylation in response to abiotic stresses in Sorghum. The Plant Journal. 2020 Jun;102(5):916-930. DOI: 10.1111/tpj.14671

[119] Wang T, Wang H, Cai D, Gao Y, Zhang H, Wang Y, Lin C, Ma L, Gu L. Comprehensive profiling of rhizome-associated alternative splicing and alternative polyadenylation in moso bamboo (Phyllostachys edulis). The Plant Journal. 2017 Aug;91(4):684-699. DOI: 10.1111/tpj.13597

[120] Cao J, Ye C, Hao G, Dabney-Smith C, Hunt AG, Li QQ. Root hair single cell type specific profiles of gene expression and alternative polyadenylation under cadmium stress. Frontiers in plant science. 2019 May 10;10:589. DOI: 10.3389/fpls.2019.00589

[121] Cai Y, Wan J. Competing endogenous RNA regulations in neurodegenerative disorders: current challenges and emerging insights. Frontiers in molecular neuroscience. 2018 Oct 5;11:370. DOI: 10.3389/fnmol.2018.00370

[122] He X, Guo S, Wang Y, Wang L, Shu S, Sun J. Systematic identification and analysis of heat-stress-responsive lncRNAs, circRNAs and miRNAs with associated co-expression and ceRNA networks in cucumber (Cucumis sativus L.). Physiologia plantarum. 2020 Mar;168(3):736-754. DOI: 10.1111/ppl.12997

[123] Sripathi VR, Choi Y, Gossett ZB, Stelly DM, Moss EM, Town CD, Walker LT, Sharma GC, Chan AP. Identification of microRNAs and their targets in four Gossypium species using RNA sequencing. Current Plant Biology. 2018 Sep 1;14:30-40. DOI: 10.1016/j.cpb.2018.09.008

[124] O'Brien J, Hayder H, Zayed Y, Peng C. Overview of microRNA biogenesis, mechanisms of actions, and circulation. Frontiers in endocrinology. 2018 Aug 3;9:402. DOI: 10.3389/fendo.2018.00402

[125] Kalvari I, Nawrocki EP, Ontiveros-Palacios N, Argasinska J, Lamkiewicz K, Marz M, Griffiths-Jones S, Toffano-Nioche C, Gautheret D, Weinberg Z, Rivas E. Rfam 14: expanded coverage of metagenomic, viral and microRNA families. Nucleic Acids Research. 2021 Jan 8;49(D1):D192-D200. DOI: 10.1093/nar/gkaa1047

[126] Stocks MB, Mohorianu I, Beckers M, Paicu C, Moxon S, Thody J, Dalmay T, Moulton V. The UEA sRNA Workbench (version 4.4): a comprehensive suite of tools for analyzing miRNAs and sRNAs. Bioinformatics. 2018 Oct 1;34(19):3382-3384. DOI: 10.1093/bioinformatics/bty338

[127] Sticht C, De La Torre C, Parveen A, Gretz N. miRWalk: An online resource for prediction of microRNA binding sites. PloS one. 2018 Oct 18;13(10):e0206239. DOI: 10.1371/journal.pone.0206239

[128] Xie F, Liu S, Wang J, Xuan J, Zhang X, Qu L, Zheng L, Yang J. deepBase v3. 0: expression atlas and interactive analysis of ncRNAs from thousands of deep-sequencing data. Nucleic acids research. 2021 Jan 8;49(D1):D877-83. DOI: 10.1093/nar/gkaa1039

[129] Vitsios DM, Kentepozidou E, Quintais L, Benito-Gutiérrez E, van Dongen S, Davis MP, Enright AJ. Mirnovo: genome-free prediction of microRNAs from small RNA sequencing data and single-cells using decision forests. Nucleic acids research. 2017 Dec 1;45(21):e177-. DOI: 10.1093/nar/gkx836

[130] Tokar T, Pastrello C, Rossos AE, Abovsky M, Hauschild AC, Tsay M, Lu R, Jurisica I. mirDIP 4.1—integrative database of human microRNA target predictions. Nucleic acids research. 2018 Jan 4;46(D1):D360-D370. DOI: 10.1093/nar/gkx1144

[131] Chen Y, Wang X. miRDB: an online database for prediction of functional microRNA targets. Nucleic acids research. 2020 Jan 8;48(D1):D127-D131. DOI: 10.1093/nar/gkz757

[132] Jha A, Shankar R. miReader: Discovering novel miRNAs in species without sequenced genome. PloS one. 2013 Jun 21;8(6):e66857. DOI: 10.1371/journal.pone.0066857

[133] Dai X, Zhuang Z, Zhao PX. psRNATarget: a plant small RNA target analysis server (2017 release). Nucleic acids research. 2018 Jul 2;46(W1):W49-W54. DOI: 10.1093/nar/gkr319

[134] Kozomara A, Birgaoanu M, Griffiths-Jones S. miRBase: from microRNA sequences to function. Nucleic acids research. 2019 Jan 8;47(D1):D155-D162. DOI: 10.1093/nar/gky1141

[135] An J, Lai J, Lehman ML, Nelson CC. miRDeep*: an integrated application

tool for miRNA identification from RNA sequencing data. Nucleic acids research. 2013 Jan 1;41(2):727-737. DOI: 10.1093/nar/gks1187

[136] Agarwal V, Bell GW, Nam JW, Bartel DP. Predicting effective microRNA target sites in mammalian mRNAs. elife. 2015 Aug 12;4:e05005. DOI: 10.7554/eLife.05005

[137] Zhao Y, Li H, Fang S, Kang Y, Wu W, Hao Y, Li Z, Bu D, Sun N, Zhang MQ, Chen R. NONCODE 2016: an informative and valuable data source of long non-coding RNAs. Nucleic acids research. 2016 Jan 4;44(D1):D203-D208. DOI: 10.1093/nar/gkv1252

[138] Ronen R, Gan I, Modai S, Sukacheov A, Dror G, Halperin E, Shomron N. miRNAkey: a software for microRNA deep sequencing analysis. Bioinformatics. 2010 Oct 15;26(20):2615-2616. DOI: 10.1093/bioinformatics/btq493

[139] Heikkinen L, Kolehmainen M, Wong G. Prediction of microRNA targets in Caenorhabditis elegans using a self-organizing map. Bioinformatics. 2011 May 1;27(9):1247-1254. DOI: 10.1093/bioinformatics/btr144

[140] Langdon WB, Petke J, Lorenz R. Evolving better RNAfold structure prediction. InEuropean Conference on Genetic Programming 2018 Apr 4 (pp. 220-236). Springer, Cham. DOI: 10.1007/978-3-319-77553-1_14

[141] Zuker Mfold©: RNA modeling program. GERF Bulletin of Biosciences. 2010.

[142] Zhang Z, Teotia S, Tang J, Tang G. Perspectives on microRNAs and phased small interfering RNAs in maize (Zea mays L.): functions and big impact on agronomic traits enhancement. Plants. 2019 Jun;8(6):170. DOI: 10.3390/plants8060170

[143] Pegler JL, Oultram JM, Grof CP, Eamens AL. Profiling the abiotic stress responsive microRNA landscape of Arabidopsis thaliana. Plants. 2019 Mar;8(3):58. DOI: 10.3390/plants8030058

[144] Ayubov MS, Mirzakhmedov MH, Sripathi VR, Buriev ZT, Ubaydullaeva KA, Usmonov DE, Norboboyeva RB, Emani C, Kumpatla SP, Abdurakhmonov IY. Role of MicroRNAs and small RNAs in regulation of developmental processes and agronomic traits in Gossypium species. Genomics. 2019 Sep 1;111(5):1018-1025. DOI: 10.1016/j.ygeno.2018.07.012

[145] Otsuka K, Yamamoto Y, Matsuoka R, Ochiya T. Maintaining good miRNAs in the body keeps the doctor away?: Perspectives on the relationship between food-derived natural products and microRNAs in relation to exosomes/extracellular vesicles. Molecular nutrition & food research. 2018 Jan;62(1):1700080. DOI: 10.1002/mnfr.201700080

[146] Barrett SP, Salzman J. Circular RNAs: analysis, expression and potential functions. Development. 2016 Jun 1;143(11):1838-1847. DOI: 10.1242/dev.128074

[147] Jeck WR, Sorrentino JA, Wang K, Slevin MK, Burd CE, Liu J, Marzluff WF, Sharpless NE. Circular RNAs are abundant, conserved, and associated with ALU repeats. Rna. 2013 Feb 1;19(2):141-157. DOI: 10.1261/rna.035667.112

[148] Vo JN, Cieslik M, Zhang Y, Shukla S, Xiao L, Zhang Y, Wu YM, Dhanasekaran SM, Engelke CG, Cao X, Robinson DR. The landscape of circular RNA in cancer. Cell. 2019 Feb 7;176(4):869-881. DOI: 10.1016/j.cell.2018.12.021

[149] Huang A, Zheng H, Wu Z, Chen M, Huang Y. Circular RNA-protein

interactions: functions, mechanisms, and identification. Theranostics. 2020;10(8):3503. DOI: 10.7150/thno.42174

[150] Hansen TB, Venø MT, Damgaard CK, Kjems J. Comparison of circular RNA prediction tools. Nucleic acids research. 2016 Apr 7;44(6):e58-. DOI: 10.1093/nar/gkv1458

[151] Liu M, Wang Q, Shen J, Yang BB, Ding X. Circbank: a comprehensive database for circRNA with standard nomenclature. RNA biology. 2019 Jul 3;16(7):899-905. DOI: 10.1080/15476286.2019.1600395

[152] Aufiero S, Reckman YJ, Tijsen AJ, Pinto YM, Creemers EE. circRNAprofiler: an R-based computational framework for the downstream analysis of circular RNAs. BMC bioinformatics. 2020 Dec;21:1-9. DOI: 10.1186/s12859-020-3500-3

[153] Li S, Li Y, Chen B, Zhao J, Yu S, Tang Y, Zheng Q, Li Y, Wang P, He X, Huang S. exoRBase: a database of circRNA, lncRNA and mRNA in human blood exosomes. Nucleic acids research. 2018 Jan 4;46(D1):D106-D112. DOI: 10.1093/nar/gkx891

[154] Zhang P, Liu Y, Chen H, Meng X, Xue J, Chen K, Chen M. CircPlant: An Integrated Tool for circRNA Detection and Functional Prediction in Plants. Genomics, Proteomics & Bioinformatics. 2020 Jun 1;18(3):352-358. DOI: 10.1016/j.gpb.2020.10.001

[155] Chu Q, Zhang X, Zhu X, Liu C, Mao L, Ye C, Zhu QH, Fan L. PlantcircBase: a database for plant circular RNAs. Molecular plant. 2017 Aug 7;10(8):1126-1128. DOI: 10.1016/j.molp.2017.03.003

[156] Sun P, Li G. CircCode: a powerful tool for identifying circRNA coding ability. Frontiers in genetics. 2019 Oct 10;10:981. DOI: 10.3389/fgene.2019.00981

[157] Chen X, Han P, Zhou T, Guo X, Song X, Li Y. circRNADb: a comprehensive database for human circular RNAs with protein-coding annotations. Scientific reports. 2016 Oct 11;6(1):1-6. DOI: 10.1038/srep34985

[158] Li L, Bu D, Zhao Y. Circ RNA wrap–a flexible pipeline for circ RNA identification, transcript prediction, and abundance estimation. FEBS letters. 2019 Jun;593(11):1179-1189. DOI: 10.1002/1873-3468.13423

[159] Glažar P, Papavasileiou P, Rajewsky N. circBase: a database for circular RNAs. Rna. 2014 Nov 1;20(11):1666-1670. DOI: 10.1261/rna.043687.113

[160] Jakobi T, Uvarovskii A, Dieterich C. circtools—a one-stop software solution for circular RNA research. Bioinformatics. 2019 Jul 1;35(13):2326-2328. DOI: 10.1093/bioinformatics/bty948

[161] Lu T, Cui L, Zhou Y, Zhu C, Fan D, Gong H, Zhao Q, Zhou C, Zhao Y, Lu D, Luo J. Transcriptome-wide investigation of circular RNAs in rice. Rna. 2015 Dec 1;21(12):2076-2087. 10.1261/rna.052282.115

[162] Zhao W, Chu S, Jiao Y. Present scenario of circular RNAs (circRNAs) in plants. Frontiers in plant science. 2019 Apr 2;10:379. DOI: 10.3389/fpls.2019.00379

[163] Wang Y, Wang Q, Gao L, Zhu B, Luo Y, Deng Z, Zuo J. Integrative analysis of circRNAs acting as ceRNAs involved in ethylene pathway in tomato. Physiologia plantarum. 2017 Nov;161(3):311-321. DOI: 10.1111/ppl.12600

[164] Li A, Huang W, Zhang X, Xie L, Miao X. Identification and characterization of CircRNAs of two pig breeds as a new biomarker in metabolism-related diseases. Cellular

Physiology and Biochemistry. 2018;47(6):2458-2470. DOI: 10.1159/000491619

[165] Zhang C, Wu H, Wang Y, Zhu S, Liu J, Fang X, Chen H. Circular RNA of cattle casein genes are highly expressed in bovine mammary gland. Journal of dairy science. 2016 Jun 1;99(6):4750-4760. DOI: 10.3168/jds.2015-10381

[166] Tang F, Barbacioru C, Wang Y, Nordman E, Lee C, Xu N, Wang X, Bodeau J, Tuch BB, Siddiqui A, Lao K. mRNA-Seq whole-transcriptome analysis of a single cell. Nature methods. 2009 May;6(5):377-382. DOI: 10.1038/nmeth.1315

[167] Saliba AE, Westermann AJ, Gorski SA, Vogel J. Single-cell RNA-seq: advances and future challenges. Nucleic acids research. 2014 Aug 18;42(14):8845-8860. DOI: 10.1093/nar/gku555

[168] Shalek AK, Benson M. Single-cell analyses to tailor treatments. Science translational medicine. 2017 Sep 20;9(408). DOI: 10.1126/scitranslmed.aan4730

[169] Chen G, Ning B, Shi T. Single-cell RNA-Seq technologies and related computational data analysis. Frontiers in genetics. 2019 Apr 5;10:317. DOI: 10.3389/fgene.2019.00317

[170] Baran-Gale J, Chandra T, Kirschner K. Experimental design for single-cell RNA sequencing. Briefings in functional genomics. 2018 Jul;17(4):233-239. DOI: 10.1093/bfgp/elx035

[171] Zhao T, Lyu S, Lu G, Juan L, Zeng X, Wei Z, Hao J, Peng J. SC2disease: a manually curated database of single-cell transcriptome for human diseases. Nucleic Acids Research. 2021 Jan 8;49(D1):D1413-D1419. DOI: 10.1093/nar/gkaa838

[172] Sokolowski DJ, Faykoo-Martinez M, Erdman L, Hou H, Chan C, Zhu H, Holmes MM, Goldenberg A, Wilson MD. Single-cell mapper (scMappR): using scRNA-seq to infer the cell-type specificities of differentially expressed genes. NAR Genomics and Bioinformatics. 2021 Mar;3(1):lqab011. DOI: 10.1093/nargab/lqab011

[173] Zhu X, Yunits B, Wolfgruber T, Liu Y, Huang Q, Poirion O, Arisdakessian C, Zhao T, Garmire D, Garmire L. GranatumX: A community engaging and flexible software environment for single-cell analysis. bioRxiv. 2019 Jan 1:385591. DOI: 10.1101/385591

[174] Svensson V, da Veiga Beltrame E, Pachter L. A curated database reveals trends in single-cell transcriptomics. Database. 2020 Jan 1;2020. DOI: 10.1093/database/baaa073

[175] Bernstein MN, Ni Z, Collins M, Burkard ME, Kendziorski C, Stewart R. CHARTS: a web application for characterizing and comparing tumor subpopulations in publicly available single-cell RNA-seq data sets. BMC bioinformatics. 2021 Dec;22(1):1-9. DOI: 10.1186/s12859-021-04021-x

[176] Li B, Gould J, Yang Y, Sarkizova S, Tabaka M, Ashenberg O, Rosen Y, Slyper M, Kowalczyk MS, Villani AC, Tickle T. Cumulus provides cloud-based data analysis for large-scale single-cell and single-nucleus RNA-seq. Nature Methods. 2020 Aug;17(8):793-798. DOI: 10.1038/s41592-020-0905-x

[177] Franzén O, Gan LM, Björkegren JL. PanglaoDB: a web server for exploration of mouse and human single-cell RNA sequencing data. Database. 2019 Jan 1;2019. DOI: 10.1093/database/baz046

[178] Franzén O, Björkegren JL. alona: a web server for single-cell RNA-seq analysis. Bioinformatics. 2020 Jun 1;36(12):3910. DOI: 10.1093/bioinformatics/btaa269

[179] Obermayer B, Holtgrewe M, Nieminen M, Messerschmidt C, Beule D. SCelVis: exploratory single cell data analysis on the desktop and in the cloud. PeerJ. 2020 Feb 19;8:e8607. DOI: 10.7717/peerj.8607

[180] Zappia L, Phipson B, Oshlack A. Exploring the single-cell RNA-seq analysis landscape with the scRNA-tools database. PLoS computational biology. 2018 Jun 25;14(6):e1006245. DOI: 10.1371/journal.pcbi.1006245

[181] Tan Y, Cahan P. SingleCellNet: a computational tool to classify single cell RNA-Seq data across platforms and across species. Cell systems. 2019 Aug 28;9(2):207-213. DOI: 10.1016/j.cels.2019.06.004

[182] Ma X, Denyer T, Timmermans MC. PscB: A Browser to Explore Plant Single Cell RNA-Sequencing Data Sets. Plant physiology. 2020 Jun 1;183(2):464-467. DOI: 10.1104/pp.20.00250

[183] Cao Y, Zhu J, Jia P, Zhao Z. scRNASeqDB: a database for RNA-Seq based gene expression profiles in human single cells. Genes. 2017 Dec;8(12):368. DOI: 10.3390/genes8120368

[184] Feng D, Whitehurst CE, Shan D, Hill JD, Yue YG. Single Cell Explorer, collaboration-driven tools to leverage large-scale single cell RNA-seq data. BMC genomics. 2019 Dec;20(1):1-8. DOI: 10.1186/s12864-019-6053-y

[185] Yang A, Troup M, Lin P, Ho JW. Falco: a quick and flexible single-cell RNA-seq processing framework on the cloud. Bioinformatics. 2017 Mar 1;33(5):767-769. DOI: 10.1093/bioinformatics/btw732

[186] Tang W, Tang AY. Biological significance of RNA-seq and single-cell genomic research in woody plants. Journal of Forestry Research. 2019 Oct;30(5):1555-1568. DOI: 10.1007/s11676-019-00933-w

[187] Tripathi RK, Wilkins O. Single cell gene regulatory networks in plants: opportunities for enhancing climate change stress resilience. Plant, Cell & Environment. 2021 Feb 1. DOI: 10.1111/pce.14012

[188] Li J, Xing S, Zhao G, Zheng M, Yang X, Sun J, Wen J, Liu R. Identification of diverse cell populations in skeletal muscles and biomarkers for intramuscular fat of chicken by single-cell RNA sequencing. BMC genomics. 2020 Dec;21(1):1-1. DOI: 10.1186/s12864-020-07136-2

[189] Foster S, Teo YV, Neretti N, Oulhen N, Wessel GM. Single cell RNA-seq in the sea urchin embryo show marked cell-type specificity in the Delta/Notch pathway. Molecular reproduction and development. 2019 Aug;86(8):931-934. DOI: 10.1002/mrd.23181

[190] Gu F, Wu J, Zhu S, Valencak TG, Liu JX, Sun HZ. Single-cell RNA-Sequencing Reveals Novel Myofibroblasts with Epithelial Cell-Like Features in the Mammary Gland of Dairy Cattle. 2020. DOI: 10.21203/rs.3.rs-101174/v1

[191] Aguiar-Pulido V, Huang W, Suarez-Ulloa V, Cickovski T, Mathee K, Narasimhan G. Metagenomics, metatranscriptomics, and metabolomics approaches for microbiome analysis: supplementary issue: bioinformatics methods and applications for big metagenomics data. Evolutionary Bioinformatics. 2016 Jan;12:EBO-S36436. DOI: 10.4137/EBO.S36436

[192] Vanwonterghem I, Jensen PD, Ho DP, Batstone DJ, Tyson GW. Linking microbial community structure, interactions and function in anaerobic digesters using new molecular techniques. Current opinion in biotechnology. 2014 Jun 1;27:55-64. DOI: 10.1016/j.copbio.2013.11.004

[193] Shakya M, Lo CC, Chain PS. Advances and challenges in metatranscriptomic analysis. Frontiers in genetics. 2019 Sep 25;10:904. DOI: 10.3389/fgene.2019.00904

[194] Jiang Y, Xiong X, Danska J, Parkinson J. Metatranscriptomic analysis of diverse microbial communities reveals core metabolic pathways and microbiome-specific functionality. Microbiome. 2016 Dec;4(1):1-8. DOI: 10.1186/s40168-015-0146-x

[195] Peimbert M, Alcaraz LD. A hitchhiker's guide to metatranscriptomics. InField Guidelines for Genetic Experimental Designs in High-Throughput Sequencing 2016 (pp. 313-342). Springer, Cham. DOI: 10.1007/978-3-319-31350-4_13

[196] Yilmaz P, Parfrey LW, Yarza P, Gerken J, Pruesse E, Quast C, Schweer T, Peplies J, Ludwig W, Glöckner FO. The SILVA and "all-species living tree project (LTP)" taxonomic frameworks. Nucleic acids research. 2014 Jan 1;42(D1):D643-D648. DOI: 10.1093/nar/gkt1209

[197] Bolyen E, Rideout JR, Dillon MR, Bokulich NA, Abnet CC, Al-Ghalith GA, Alexander H, Alm EJ, Arumugam M, Asnicar F, Bai Y. Reproducible, interactive, scalable and extensible microbiome data science using QIIME 2. Nature biotechnology. 2019 Aug;37(8):852-857. DOI: 10.1038/s41587-019-0209-9

[198] McDonald D, Price MN, Goodrich J, Nawrocki EP, DeSantis TZ, Probst A, Andersen GL, Knight R, Hugenholtz P. An improved Greengenes taxonomy with explicit ranks for ecological and evolutionary analyses of bacteria and archaea. The ISME journal. 2012 Mar;6(3):610-618. DOI: 10.1038/ismej.2011.139

[199] Westreich ST, Treiber ML, Mills DA, Korf I, Lemay DG. SAMSA2: a standalone metatranscriptome analysis pipeline. BMC bioinformatics. 2018 Dec;19(1):1-1. DOI: 10.1186/s12859-018-2189-z

[200] Huerta-Cepas J, Szklarczyk D, Heller D, Hernández-Plaza A, Forslund SK, Cook H, Mende DR, Letunic I, Rattei T, Jensen LJ, von Mering C. eggNOG 5.0: a hierarchical, functionally and phylogenetically annotated orthology resource based on 5090 organisms and 2502 viruses. Nucleic acids research. 2019 Jan 8;47(D1):D309-D314. DOI: 10.1093/nar/gky1085

[201] Batut B, Gravouil K, Defois C, Hiltemann S, Brugère JF, Peyretaillade E, Peyret P. ASaiM: a Galaxy-based framework to analyze microbiota data. GigaScience. 2018 Jun;7(6):giy057. DOI: 10.1093/gigascience/giy057

[202] Tatusova T, Ciufo S, Fedorov B, O'Neill K, Tolstoy I. RefSeq microbial genomes database: new representation and annotation strategy. Nucleic acids research. 2014 Jan 1;42(D1):D553-D559. DOI: 10.1093/nar/gkt1274

[203] Wilke A, Bischof J, Gerlach W, Glass E, Harrison T, Keegan KP, Paczian T, Trimble WL, Bagchi S, Grama A, Chaterji S. The MG-RAST metagenomics database and portal in 2015. Nucleic acids research. 2016 Jan 4;44(D1):D590-D594. DOI: 10.1093/nar/gkv1322

[204] Overbeek R, Olson R, Pusch GD, Olsen GJ, Davis JJ, Disz T, Edwards RA, Gerdes S, Parrello B, Shukla M, Vonstein V. The SEED and the Rapid Annotation of microbial genomes using Subsystems Technology (RAST). Nucleic acids research. 2014 Jan 1;42(D1):D206-D214. DOI: 10.1093/nar/gkt1226

[205] Martinez X, Pozuelo M, Pascal V, Campos D, Gut I, Gut M, Azpiroz F,

Guarner F, Manichanh C. MetaTrans: an open-source pipeline for metatranscriptomics. Scientific reports. 2016 May 23;6(1):1-2. DOI: 10.1038/srep26447

[206] Singh DP, Prabha R, Gupta VK, Verma MK. Metatranscriptome analysis deciphers multifunctional genes and enzymes linked with the degradation of aromatic compounds and pesticides in the wheat rhizosphere. Frontiers in microbiology. 2018 Jul 3;9:1331. DOI: 10.3389/fmicb.2018.01331

[207] Li F. Metatranscriptomic profiling reveals linkages between the active rumen microbiome and feed efficiency in beef cattle. Applied and environmental microbiology. 2017 May 1;83(9). DOI: 10.1128/AEM.00061-17

[208] Song Z, Du H, Zhang Y, Xu Y. Unraveling core functional microbiota in traditional solid-state fermentation by high-throughput amplicons and metatranscriptomics sequencing. Frontiers in microbiology. 2017 Jul 14;8:1294. DOI: 10.3389/fmicb.2017.01294

[209] Weckx S, Van der Meulen R, Allemeersch J, Huys G, Vandamme P, Van Hummelen P, De Vuyst L. Community dynamics of bacteria in sourdough fermentations as revealed by their metatranscriptome. Applied and environmental microbiology. 2010 Aug 15;76(16):5402-5408. DOI: 10.1128/AEM.00570-10

[210] Peng J, Wegner CE, Bei Q, Liu P, Liesack W. Metatranscriptomics reveals a differential temperature effect on the structural and functional organization of the anaerobic food web in rice field soil. Microbiome. 2018 Dec;6(1):1-6. DOI: DOI: 10.1186/s40168-018-0546-9

[211] Kukurba KR, Montgomery SB. RNA sequencing and analysis: Cold Spring Harbor Protocols. 2015;Nov 1;2015(11):pdb-top084970. DOI: 10.1101/pdb.top084970

[212] Khang TF, Lau CY. Getting the most out of RNA-seq data analysis. PeerJ. 2015 Oct 29;3:e1360. DOI: 10.7717/peerj.1360

[213] Khosravi P, Gazestani VH, Pirhaji L, Law B, Sadeghi M, Goliaei B, Bader GD. Inferring interaction type in gene regulatory networks using co-expression data. Algorithms for molecular biology. 2015 Dec;10(1):1-1. DOI: 10.1186/s13015-015-0054-4

[214] Han Y, Gao S, Muegge K, Zhang W, Zhou B. Advanced applications of RNA sequencing and challenges. Bioinformatics and biology insights. 2015 Jan;9:BBI-S28991. DOI: 10.4137%2FBBI.S28991

[215] Delgado FM, Gómez-Vela F. Computational methods for Gene Regulatory Networks reconstruction and analysis: A review. Artificial intelligence in medicine. 2019 Apr 1;95:133-145. DOI: 10.1016/j.artmed.2018.10.006

[216] Macho Rendón J, Lang B, Ramos Llorens M, Gaetano Tartaglia G, Torrent Burgas M. DualSeqDB: the host–pathogen dual RNA sequencing database for infection processes. Nucleic Acids Research. 2021 Jan 8;49(D1):D687-D693. DOI: 10.1093/nar/gkaa890

[217] Sebastian S, Ali SA, Das A, Roy S. pARACNE: A Parallel Inference Platform for Gene Regulatory Network Using ARACNe. InInnovations in Computational Intelligence and Computer Vision 2021 (pp. 85-92). Springer, Singapore. DOI: 10.1007/978-981-15-6067-5_11

[218] Arkin AP, Cottingham RW, Henry CS, Harris NL, Stevens RL, Maslov S, Dehal P, Ware D, Perez F, Canon S, Sneddon MW. KBase: the United States department of energy systems biology knowledgebase. Nature biotechnology. 2018 Aug;36(7):566-569. DOI: 10.1038/nbt.4163

[219] Van de Sande B, Flerin C, Davie K, De Waegeneer M, Hulselmans G, Aibar S, Seurinck R, Saelens W, Cannoodt R, Rouchon Q, Verbeiren T. A scalable SCENIC workflow for single-cell gene regulatory network analysis. Nature Protocols. 2020 Jul;15(7):2247-2276. DOI: 10.1038/s41596-020-0336-2

[220] Boccaletto P, Machnicka MA, Purta E, Piątkowski P, Bagiński B, Wirecki TK, de Crécy-Lagard V, Ross R, Limbach PA, Kotter A, Helm M. MODOMICS: a database of RNA modification pathways. 2017 update. Nucleic acids research. 2018 Jan 4;46(D1):D303-D307. DOI: 10.1093/nar/gkx1030

[221] Dibaeinia P, Sinha S. SERGIO: a single-cell expression simulator guided by gene regulatory networks. Cell Systems. 2020 Sep 23;11(3):252-271. DOI: 10.1016/j.cels.2020.08.003

[222] Keseler IM, Mackie A, Peralta-Gil M, Santos-Zavaleta A, Gama-Castro S, Bonavides-Martínez C, Fulcher C, Huerta AM, Kothari A, Krummenacker M, Latendresse M. EcoCyc: fusing model organism databases with systems biology. Nucleic acids research. 2013 Jan 1;41(D1):D605-D612. DOI: 10.1093/nar/gks1027

[223] Moerman T, Aibar Santos S, Bravo González-Blas C, Simm J, Moreau Y, Aerts J, Aerts S. GRNBoost2 and Arboreto: efficient and scalable inference of gene regulatory networks. Bioinformatics. 2019 Jun 1;35(12):2159-2161. DOI: 10.1093/bioinformatics/bty916

[224] Anders G, Mackowiak SD, Jens M, Maaskola J, Kuntzagk A, Rajewsky N, Landthaler M, Dieterich C. doRiNA: a database of RNA interactions in post-transcriptional regulation. Nucleic acids research. 2012 Jan 1;40(D1):D180-D186. DOI: 10.1093/nar/gkr1007

[225] Geurts P. dynGENIE3: dynamical GENIE3 for the inference of gene networks from time series expression data. Scientific reports. 2018 Feb 21;8(1):1-2. DOI: 10.1038/s41598-018-21715-0

[226] Musungu B, Bhatnagar D, Quiniou S, Brown RL, Payne GA, O'Brian G, Fakhoury AM, Geisler M. Use of Dual RNA-seq for Systems Biology Analysis of Zea mays and Aspergillus flavus interaction. Frontiers in Microbiology. 2020 Jun 3;11:853. DOI: 10.3389/fmicb.2020.00853

[227] D'Esposito D, Ferriello F, Dal Molin A, Diretto G, Sacco A, Minio A, Barone A, Di Monaco R, Cavella S, Tardella L, Giuliano G. Unraveling the complexity of transcriptomic, metabolomic and quality environmental response of tomato fruit. BMC plant biology. 2017 Dec;17(1):1-8. DOI: 10.1186/s12870-017-1008-4

[228] Rodenburg SY, Seidl MF, De Ridder D, Govers F. Genome-wide characterization of Phytophthora infestans metabolism: a systems biology approach. Molecular plant pathology. 2018 Jun;19(6):1403-1413. DOI: 10.1111/mpp.12623

[229] Croote D, Quake SR. Food allergen detection by mass spectrometry: the role of systems biology. NPJ systems biology and applications. 2016 Sep 29;2(1):1-0. DOI: 10.1038/npjsba.2016.22

[230] Gao Z, Ding R, Zhai X, Wang Y, Chen Y, Yang CX, Du ZQ. Common Gene Modules Identified for Chicken Adiposity by Network Construction and Comparison. Frontiers in genetics. 2020 May 29;11:537. DOI: 10.3389/fgene.2020.00537